Lead Free Solders

Edited by Abhijit Kar

Published in London, United Kingdom

IntechOpen

Supporting open minds since 2005

Lead Free Solders
http://dx.doi.org/10.5772/intechopen.76955
Edited by Abhijit Kar

Assistant to the Editor(s): Monalisa Char

Contributors
Shuye Zhang, Tiesong Lin, Peng He, Kyung-Wook Paik, Roman Koleňák, Martin Provazník, Igor Kostolný, Mitsuhiro Watanabe, Eiichi Kondoh, Anantha Padmanaban D, Arun Vasantha Geethan, Abhijit Kar, Monalisa Char

Notice
Statements and opinions expressed in the chapters are these of the individual contributors and not necessarily those of the editors or publisher. No responsibility is accepted for the accuracy of information contained in the published chapters. The publisher assumes no responsibility for any damage or injury to persons or property arising out of the use of any materials, instructions, methods or ideas contained in the book.

First published in London, United Kingdom, 2019 by IntechOpen
IntechOpen is the global imprint of INTECHOPEN LIMITED, registered in England and Wales, registration number: 11086078, The Shard, 25th floor, 32 London Bridge Street London, SE19SG – United Kingdom
Printed in Croatia

British Library Cataloguing-in-Publication Data
A catalogue record for this book is available from the British Library

Additional hard and PDF copies can be obtained from orders@intechopen.com

Lead Free Solders
Edited by Abhijit Kar
p. cm.
Print ISBN 978-1-78985-459-6
Online ISBN 978-1-78985-460-2
eBook (PDF) ISBN 978-1-83962-278-6

We are IntechOpen,
the world's leading publisher of
Open Access books
Built by scientists, for scientists

4,300+
Open access books available

116,000+
International authors and editors

130M+
Downloads

Our authors are among the

151
Countries delivered to

Top 1%
most cited scientists

12.2%
Contributors from top 500 universities

Interested in publishing with us?
Contact book.department@intechopen.com

Numbers displayed above are based on latest data collected.
For more information visit www.intechopen.com

Meet the editor

Abhijit Kar received his PhD in Chemistry-Materials Science from Jadavpur University, India. Dr. Kar carried out his postdoctoral research in different parts of the world, e.g. Sungkyunkwan University, South Korea, Swiss Federal Laboratory at Zürich, Switzerland, and as a visiting scientist at Rühr Universität, Germany. Dr. Kar has about 17 years of research and teaching experience. His current research interest comprises the application of nanotechnology for advanced materials. He has worked on a number of similar and dissimilar materials joining methods and techniques. He is one of the first research scientists to work on lead-free solder for electronics/microelectronics applications in India. He has developed expertise on different materials characterization techniques and the mechanical and functional property evaluation of materials. Dr. Kar has published around 40 research papers in journals, edited the book *Nanoelectronics and Materials Development*, and contributed a book chapter on electron microscopy. He serves as a reviewer of many international peer-reviewed journals from Elsevier and Springer. He is editorial board member of *Journal of Materials Sciences and Applications*, *American Association for Science and Technology*, *Journal of Energy and Natural Resources*, etc.

Contents

Preface

This book deals with some of the most advanced research observations and their applications in electronic device manufacturing in a very concise and structured way. It will help readers to gain in-depth knowledge about lead-free solder and its significance. Generally, the electronics world has been shifting towards lead-free electronic devices because of strict rules, government legislation, and serious health and environmental issues arising out of the use of lead in electronic devices.

It is now a great challenge for academia and industry to study and analyze lead-free solder joints and to recommend the manufacturing of reliable lead-free electronic devices. Moreover, the task becomes more complicated when miniaturization hit the electronics world recently.

All five chapters were written by acclaimed researchers from all over the world describing detailed and necessary facts and figures on all aspects and scopes of lead-free soldering, its application, and the immense possibilities of advanced materials development.

Chapter 1 gives a general overview and discusses the need for lead-free solder research. It also proposes some of the very recent observations towards achieving reliable lead-free solder joints. Kar et al. observe very interesting effects of using multilayers of Sn for controlling intermetallic compounds growth at the lead-free solder join interface. *Chapter 2* describes the importance of intermetallics in lead-based and Pb-free solder alloys starting from their formation, characteristics, and effects on material joints. Since the growth of intermetallics over time is determined by solid-state diffusion rules and varies from system to system, readers will get a broad overview of the interaction of intermetallics with substrates and will be able to determine the exact effect on solder properties. The chapter explains how the mechanical properties and growth kinetics of intermetallics at the solder joint interface can be altered by the addition of rare earth elements. *Chapter 3* discusses the soldering of metallic and ceramic materials by lead-free active Sn and Bi-In-based solders. Readers will gain detailed knowledge of the possibilities of soldering ceramic materials, which is limited due to the poor wettability of ceramic substrates with commercial solders at classical soldering technologies because of the different thermal expansions of soldered materials. This chapter describes how the technology of soldering with active solders is selected for joining ceramic materials along with applications of ultrasound for mechanical activation of solders, which disrupts the surface oxides, changes the surface energy of ceramic materials, and supports the diffusion processes in the interface. *Chapter 4* explains the room temperature formation of intermixing layers between a Cu/glass stack. For fabricating highly reliable Cu/glass structures, atomically scaled interface bonding is currently necessary. This chapter describes how these nanoscale ZnO adhesion layers are prepared and used for reducing the processing temperature so that it does not affect the packaging technologies. *Chapter 5* discusses ways of improving lead-free micron-sized solder joint cracks. These can be used in flex-on-board interconnections, which have become very popular in mobile electronics applications. However, crack formation within these micron-sized solder joints remains a major problem, which occurs in low-temperature curable acrylic polymer resins after bonding processes. The authors describe how the elastic modulus of an anisotropic conductive film resin assembly can be altered for low-melting solder materials and electronic device packaging so that these cracks can be controlled and reliability maintained. In this study, the mechanism of an Sn-58Bi solder joint crack with low-temperature curable acrylic adhesive is investigated.

The main focus of this book is to make scientists/researchers aware of the most advanced developments and applications in lead-free solders. We hope this book will be helpful not only for those who work in the field of solder jointing and its applications but also for researchers and scientists working on photonics and semiconductor devices. Finally, we would like to thank all the contributing authors for their hard work in the preparation of this book.

Abhijit Kar
Jagadis Bose National Science Talent Search,
Kolkata, India

Introductory Chapter: Overview of Pb-Free Solders and Effect of Multilayered Thin Film of Sn on the Lead-Free Solder Joint Interface

Monalisa Char and Abhijit Kar

1. Introduction

All electronic devices starting from a small remote-controlled car to large aerospace vehicles require numerous interconnects within the circuit to complete the electrical pathways for its smooth functionality. For making these contact points, soldering plays the most important role. Until recently, Sn63-Pb37 (Sn-Pb eutectic) solders were used widely for making these contact points. The whole electronic industry was depending on the Sn-Pb alloy due to their low cost, good solder ability, low melting temperature, and satisfactory mechanical and functional properties. However, it has been observed that there arise serious environmental and health hazards caused due to the extensive use of electronic gadgets containing lead. So, there came strict restrictions imposed by the Restriction of Hazardous Substance (2002/95/EC (RoHS 1)) and Waste Electrical and Electronic (WEEE) directives and other similar bodies in use of lead in the electronic gadgets. Hence, the need for the development of lead-free solder alloy for electronics and microelectronic devices has come into the limelight.

However, the miniaturization of electronic devices aiming at high performance has made the task very complex for all the researchers to develop reliable and cost-effective electronic joining materials/technologies. Scientists have developed different lead-free soldering alloys for making consumer electronic devices; however, after more than couples of decades, it is yet to conclude the best possible solder that can withstand different harsh environmental conditions and provide good reliability and improve the service life of the device/joints (compared to the Sn-Pb eutectic solder). In recent days, we have observed that the service life of any electronic gadgets have been reduced drastically. These early failures of various consumer electronic gadgets have been often linked with the absence of a good lead-free solder material, which can withstand various service-exposed conditions [1–8].

After many years of research, different lead-free solder alloys like SAC105, SAC205, SAC305, SAC405, SN100C, Sn-Zn-, Sn-Cu-, and Sn-Bi-based solders have been explored by the industry and academia [9–11]. However, SAC305 (Sn-3.0Ag-0.5Cu), a lead-free solder alloy, has been identified as a nearest substitute of the conventionally used Sn37Pb solder.

Although SAC305 alloy is widely used in electronic industry, there is still a large scope for its modification or introduction of other joining technologies/procedures in order to get the optimized reliable joints, which we lack even now. Joints produced using such alloys have been investigated extensively by different researchers. It has been found that the interfacial microstructure, orientation of IMC at the interface, and its volume fraction are the key factors for optimizations of the mechanical and functional properties of the joints [12–15].

It has been realized that focusing only on Pb-free solder alloy development would certainly reduce the environmental pollution. But while achieving this, rise in processing temperature (because of the higher melting point of the lead-free solder) would in turn enhance the energy consumption while manufacturing the solder joints, thereby causing thermal pollution again. So, we have focused on finding suitable method/s for lowering the processing temperature and controlling the interfacial intermetallics growth of lead-free soldering, thereby increasing the reliability of the gadgets.

In this regard for lowering the processing temperature, we have used the transient liquid phase (TLP)-like soldering technique to join two copper substrates along with solder alloy Sn-3.0Ag-0.5Cu (SAC305). In case of high-temperature metal joining (welding/brazing), it is observed that the TLP bonding along with some interlayers generally produces IMC layer at the interface with lesser thickness compared to that in the conventional brazing/welding. Also, it produces homogeneous interface [16]. By dint of TLP-like soldering technique, we have prepared lead-free solder joint of Cu using thin multilayer structure of tin along with SAC305 solder alloy, hereafter termed as Cu-Sn/SAC/Sn-Cu solder system. In order to compare the effect of this thin multilayered structure, conventional solder joint of copper pads using SAC305, hereafter termed as Cu-SAC-Cu, has been prepared. We have produced the solder joints at 230°C, which is about 15–30°C lower than that of the conventional reflow soldering process generally used for making the lead-free solder joints [17].

Already reported studies in the literature show that joints produced at high processing temperature consist of higher volume fraction of IMCs with inhomogeneous interfacial microstructure (mostly scallop shaped), which essentially produce internal stress and exhibit premature failure [18].

It is found that the interfacial morphology and distribution of the IMC across the interface plays the key role in determining the quality and reliability of any solder joints [4, 19, 20]. Therefore, researchers have attempted to develop homogeneous interfacial microstructure with restricted growth of intermetallics in order to produce solder joints with better mechanical and functional properties. But, it is yet to conclude the best suitable and optimized temperature or processing technique that would give us reliable lead-free solder joint with improved service life of the electronic gadgets [21–24].

For Sn-3.0Ag-0.5Cu/Cu solder joint prepared at 250°C, Tong et al. showed that after long hours of aging (~288 h) at 150°C, the growth rate of Cu_3Sn surpasses the growth rate of Cu_6Sn_5 phase, where the total IMC thickness increases by almost 97.56% after aging of only 500 h [25].

Compared to conventional Cu-SAC-Cu solder joints (**Figure 1**), we have been able to produce Cu-Sn/SAC/Sn-Cu solder joints at 230°C (**Figure 2**), which exhibit a reduced growth rate with respect to the total IMC thickness and show homogeneous interfacial microstructure structure across the solder joint interface. Interestingly, even after aging of 1200 h at 150°C, it has been observed that there is only 68.85% increment in the total IMC thickness for Cu-Sn/SAC/Sn-Cu solder joint (**Figure 3a**), whereas there is an increment in the total IMC thickness of about 78% for conventional Cu-SAC-Cu solder joint (**Figure 4a**). It has also been observed that the growth rate of conventional Cu-SAC-Cu solder joint is increasing rapidly from 0.005 to around 0.01 μm/h at 1200 h of aging (**Figure 4b**). Whereas a reduced growth rate of total IMC thickness has been detected for Cu-Sn/SAC/Sn-Cu solder joint across the solder joint interface, which is gradually decreasing from 0.008 (as soldered condition) to 0.001 μm/h at 1200 h of aging, suggesting that the IMC formation has been controlled with the help of thin multilayer of tin film as shown in **Figure 3b**.

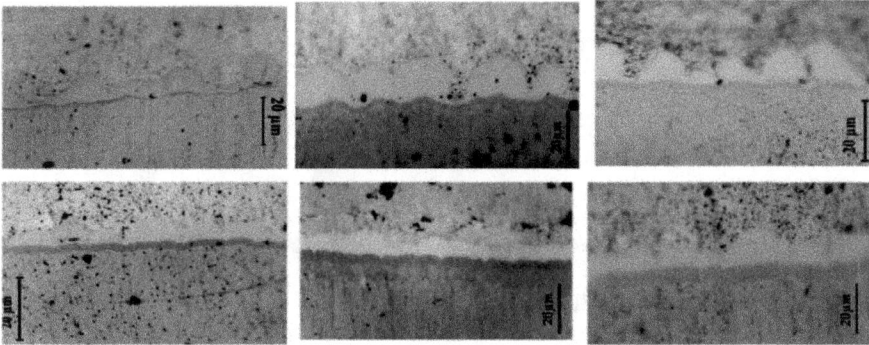

Figure 1.
Optical images of the IMC microstructure of Cu-SAC-Cu solder joint with varying aging time at 150°C. (a) As cast, (b) 100 h aging, (c) 300 h aging, (d) 500 h aging, (e) 700 h aging, and (f) 1200 h aging.

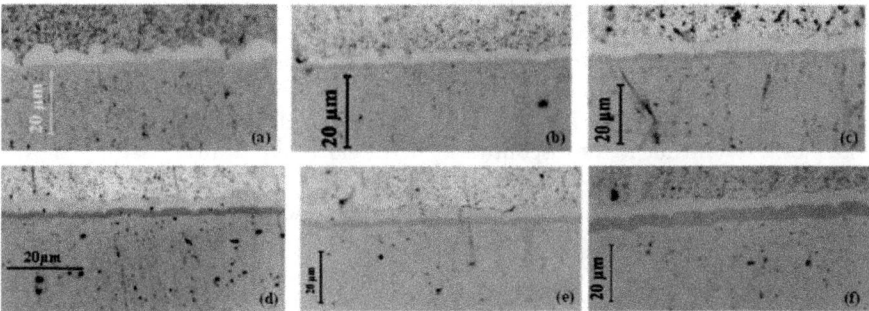

Figure 2.
Optical images of the IMC microstructure of Cu-Sn/SAC/Sn-Cu solder joint with varying aging time at 150°C. (a) As cast, (b) 100 h aging, (c) 300 h aging, (d) 500 h aging, (e) 700 h aging, and (f) 1200 h aging.

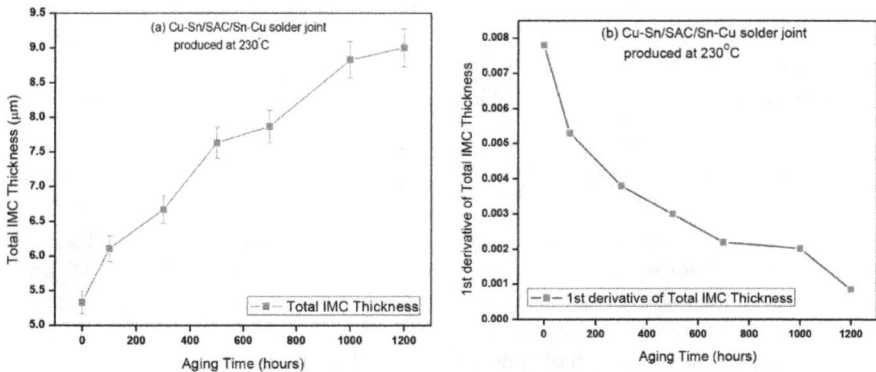

Figure 3.
(a) Total IMC thickness of Cu-Sn/SAC/Sn-Cu solder joint with variation of aging time in hours. (b) First derivative of total IMC thickness of Cu-Sn/SAC/Sn-Cu solder joint with aging time in hours.

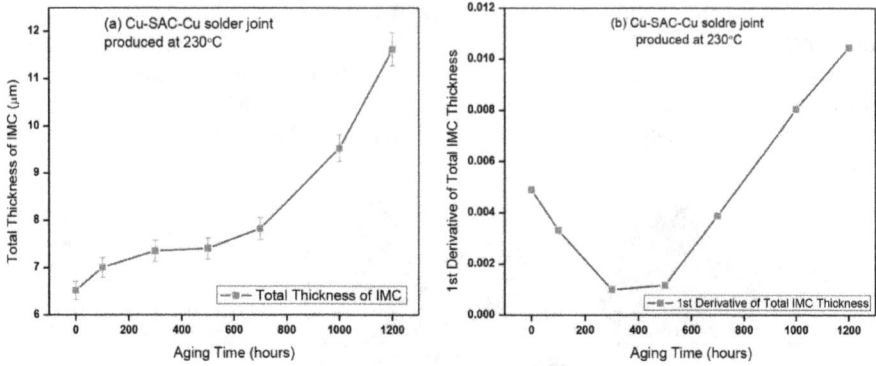

Figure 4.
(a) Plot of total IMC thickness of Cu-SAC-Cu solder joint with variation of aging time in hours.
(b) Derivative plot of total IMC thickness of Cu-SAC-Cu solder joint with variation of aging time in hours.

In order to control the IMC thickness across the solder joint interface, many researchers have also tried to add the fourth element, which would get mixed up with the solder alloy and would restrict the diffusion of Cu from substrate toward solder and thereby restricting tin to get diffused from solder toward substrate. But, an optimized growth rate of IMC formation has not been achieved yet. Ni-doped SAC solder alloy has been studied by Benabou et al. Ni is considered to be one of the best elements in restricting the diffusion of copper across the interface [26]. Study shows that though Ni-doped solder alloy produces joints with thinner IMC layer, it could not be able to restrict the formation of voids in Cu_3Sn layer with aging time and it produces cavities in $(Cu, Ni)_6Sn_5$ phase, which would be the point of initiation of failure of solder joint. Also, it increases the electrical resistivity of the solder joint system due to the discontinuity in the electrical pathways because of the presence of the voids.

Kang et al. have reported SAC solder alloy with surface finish of Ni(P)/Au, which produces Ni_3Sn_4 IMC layer, which grows faster than Cu-Sn IMC (Cu_6Sn_5 and Cu_3Sn phases) on bare Cu and remains detached from the substrate. This detachment of the Ni_3Sn_4 phase from the substrate decreases the shear strength as well as increases the electrical resistivity of the solder joint due to the discontinuity in the electrical pathways [27]. However, our proposed multilayer structure in the Cu-Sn/SAC/Sn-Cu solder joint shows neither void at the Cu_3Sn phase nor any cavities at Cu_6Sn_5 phase; moreover, a controlled rate of IMC formation has been observed. This may be attributed due to the formation of solid solution of β-Sn phase in the interface, which restricts the Cu diffusion from substrate to solder. On the other hand, a significant improvement in the electrical properties has been observed for the solder joints produced using this TLP-like soldering and multilayered structure.

2. Electrical conductivity

The electrical resistivity across the solder joints has been investigated by four probe methods. For the Cu-Sn/SAC/Sn-Cu solder joint it was found $1.4*10^{-5}$ Ω-mm for as soldered samples and $2.66*10^{-5}$ Ω-mm after 1200 h of thermal aging, whereas the electrical resistivity of the conventional Cu-SAC-Cu solder joint has been found to be $4.9*10^{-5}$ Ω-mm at 1200 h of thermal aging, which is almost twice that of Cu-Sn/SAC/Sn-Cu solder joint aged up to 1200 h. Thinner interfacial IMC layers cause lower associated localized Joule heating, resulting in enhanced electronic transport across the joints.

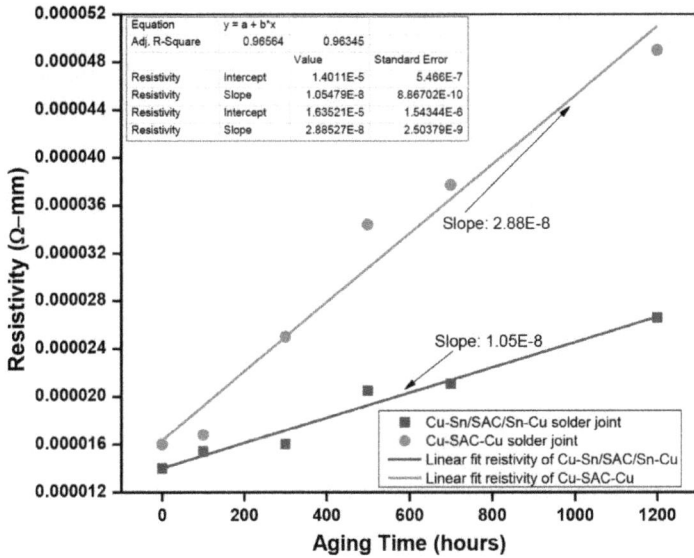

Figure 5.
Resistivity plot of Cu-Sn/SAC/Sn-Cu solder system and Cu-SAC-Cu solder joint system with varying aging time.

The rate of increment of resistivity is also slow in case of Cu-Sn/SAC/Sn-Cu solder joints than that of the Cu-SAC-Cu solder joint, as revealed from the slope of the curve (**Figure 5**). This may be attributed due to controlled IMC formation at the interface achieved by using thin multilayers of Sn film for Cu-Sn/SAC/Sn-Cu solder joints. Sn has lower resistivity than SAC solder paste (resistivity value: Sn ≈ 1.15 and SAC ≈ 3.57 $\mu\Omega$ cm). Also, ionization enthalpy of Sn is lower than that of Cu ($Cu_{Ionization\ Energy}$ = 745.5 and $Sn_{Ionization\ Energy}$ = 708.6 kJ/mol); therefore, Sn loses its outermost electron much earlier than Cu after getting little amount of energy, so it provides more number of charge carriers to conduct electricity, thereby reducing its resistance and increasing the conductivity in comparison with the Cu-SAC-Cu solder joints where no multilayers are been used.

3. Conclusion

This introductory chapter discusses the overview of the lead-free solders, its importance, and necessity in the cutting-edge research carried out across the globe for making environment friendly electronic devices. For any solder joints, the interfacial characteristics like mechanical and functional properties along with the microstructural morphology play the mostly vital role. Therefore, a systematic and detailed knowledge of the interfacial products and their structures are of fundamental interest for understanding the behavior of the solder joints. The concept of transient liquid phase-like soldering technique has been discussed, and the effect of thin multilayered film of Sn has been reported toward obtaining a more reliable lead-free solder joint.

We have demonstrated that incorporation of thin multilayer of Sn film along with SAC305 paste could produce good solder joints at 230°C. Cu pads have been joined by TLP-like soldering. For Cu-Sn/SAC/Sn-Cu solder joints, homogeneous interfacial microstructures have been formed across the solder joint interface as compared to scallop like IMC in the conventional Cu-SAC-Cu solder joint. Incorporation of thin multilayers of Sn film is capable of reducing the IMC growth rate up to 10% compared to the conventional Cu-SAC-Cu solder joint. Thinner interfacial IMC layers cause lower associated localized Joule heating, resulting in enhanced electronic transport across the joint, thereby reducing the electrical resistivity and increasing the conductivity across the solder joint interface.

Author details

Monalisa Char[1,2] and Abhijit Kar[1*]

1 Jagadis Bose National Science Talent Search, Kolkata, India

2 National Institute of Technology, Durgapur, India

*Address all correspondence to: chatrak130@yahoo.co.in

IntechOpen

References

[1] Zhang L, Wang ZG, Shang JK. Current-induced weakening of Sn3.5Ag0.7Cu Pb-free solder joints. Scripta Materialia. 2007;**56**(5):381-384

[2] Mahmood T, Mian A, Amin MR, Auner G, Witte R, Herfurth H, et al. Finite element modeling of transmission laser microjoining process. Journal of Materials Processing Technology. 2007;**186**(1):37-44

[3] Tu KN, Thompson RD. Kinetics of interfacial reaction in bimetallic Cu-Sn thin films. Acta Metallurgica. 1982;**30**(5):947-952

[4] Kar A, Ghosh M, Majumdar BS, Ghosh RN, Ray AK. Interfacial microstructure, shear strength and electrical conductivity of Sn-3.5Ag-0.5In/Cu lead free soldered joints. Materials Technology. 2007;**22**(3):161-165

[5] Tu K-N. Copper–Tin Reactions in Bulk Samples. In: Solder Joint Technology, Springer Series in Materials Science Vol. 117. New York: Springer; 2007. pp. 37-71

[6] Gao F, Takemoto T, Nishikawa H. Effects of Co and Ni addition on reactive diffusion between Sn-3.5Ag solder and Cu during soldering and annealing. Materials Science and Engineering A. 2006;**420**(1-2):39-46

[7] Xu L, Pang JHL. Nano-indentation characterization of Ni–Cu–Sn IMC layer subject to isothermal aging. Thin Solid Films. 2006;**504**(1-2):362-336

[8] Luo WC, Ho CE, Tsai JY, Lin YL, Kao CR. Solid-state reactions between Ni and Sn-Ag-Cu solders with different Cu concentrations. Materials Science and Engineering A. 2005;**396**(1):385-391

[9] Cheng S, Huang C-M, Pecht M. A review of lead free solders for electronics applications. Microelectronics Reliability. 2017;**75**:77-95

[10] Berni R, Catelani M, Fiesoli C, Scarano VL. A comparison of alloy-surface finish combinations considering different component package types and their impact on soldering reliability. IEEE Transactions on Reliability. 2016;**65**(1):272-281

[11] Coyle R, Sweatman K, Arfaei B. Thermal fatigue evaluation of Pb-free solder joints: Results, lessons learned, and future trends. Journal of the Minerals, Metals and Materials Society. 2015;**67**(10):2394-2415

[12] Kim KS, Huh SH, Suganuma K. Effects of intermetallic compounds on properties of Sn-Ag-Cu lead-free soldered joints. Journal of Alloys and Compounds. 2003;**352**(1-2):226-236

[13] Chan YC, So ACK, Lai JKL. Growth kinetic studies of Cu-Sn intermetallic compound and its effect on shear strength of LCCC SMT solder joints. Materials Science and Engineering B. 1998;**55**(1-2):5-13

[14] Li JF, Agyakwa PA, Johnson CM. Interfacial reaction in Cu/Sn/Cu system during the transient liquid phase soldering process. Acta Materialia. 2011;**59**(3):1198-1211

[15] Deng X, Piotrowski G, Williams JJ and Chawla N. "Influence of initial morphology and thickness of Cu6Sn5 and Cu3Sn intermetallics on growth and evolution during thermal aging of Sn-Ag solder/Cu joints," Journal of Electronic Materials. 2003;**12**(32):1403-1413

[16] Egar T, Sadeghian M, Ekrami A, Jamshidi R. Transient liquid phase bonding of 304 stainless steel using a Co-based interlayer. Science and Technology of Welding and Joining. 2017;**8**(12):666-672

[17] Leong YM, Haseeb ASMA. Soldering characteristics and mechanical properties of Sn-1.0Ag-0.5Cu solder with minor aluminum addition. Materials. 2016;**9**(7):1-17. DOI: 10.3390/ma9070522

[18] Tan AT, Tan AW, Yusof F. Influence of nanoparticle addition on the formation and growth of intermetallic compounds (IMCs) in Cu/Sn-Ag-Cu/Cu solder joint during different thermal conditions. Science and Technology of Advanced Materials. 2015;**16**:033505

[19] Mohammad Faizan. "Effect of inter metallic phase morphology on substrate dissolution during reflow soldering," International Journal of Mechanical Engineering and Computer Applications. 2014;**2**(2):28-32. ISSN 2320-6349

[20] Ghosh M, Kar A, Das SK, Ray AK. Aging characteristics of Sn-Ag eutectic solder alloy with the addition of Cu, In, and Mn. Metallurgical and Materials Transactions A. 2009;**40**(10):2369-2376

[21] Laurila T, Hurtig J, Vuorinen V, Kivilahti JK. Effect of Ag, Fe, Au and Ni on the growth kinetics of Sn-Cu intermetallic compound layers. Microelectronics Reliability. 2008;**49**(3):242-247 DOI: 10.1016/j.microrel.2008.08.007

[22] Chia PY, Haseeb ASMA, Mannan SH. Reactions in electrodeposited Cu/Sn and Cu/Ni/Sn nanoscale multilayers for interconnects. Materials. 2016;**9**:430. DOI: 10.3390/ma9060430

[23] Bui QV, Nam ND, Yoon JW, Choi DH, Kar A, Kim JG, et al. Effect of gold on the corrosion behavior of an electroless nickel/immersion gold surface finish. Journal of Electronic Materials. 2011;**40**(9):1937-1942

[24] Yoon J-W, Noh B-I, Kim B-K, Shur C-C, Jung S-B. Wettability and interfacial reactions of Sn–Ag–Cu/Cu and Sn–Ag–Ni/Cu solder joints. Journal of Alloys and Compounds. 2009;**486**:142-147

[25] An T, Qin F. Effects of the intermetallic compound microstructure on the tensile behavior of Sn-3.0Ag-0.5Cu/Cu solder joint under various strain rates. Microelectronics Reliability. 2014;**54**:932-938. DOI: 10.1016/j.microrel.2014.01.008

[26] Benabou L, Vivet L, Tao QB, Tran NH. Microstructural effects of isothermal aging on a doped SAC solder alloy. International Journal of Materials Research. 2018;**109**(1):76-82. DOI: 10.3139/146.111578

[27] Kang SK, Choi WK, Yim MJ, Shih DY. Studies of the mechanical and electrical properties of lead free solder joints. Journal of Electronic Materials. 2002;**11**(31):1292-1303

Chapter 2

Role of Intermetallics in Lead-Free Alloys

Dattaguru Ananthapadmanaban and Arun Vasantha Geethan

Abstract

Intermetallics are intermediate compounds formed between two metals. They are usually brittle. The presence of intermetallics leads to deterioration in mechanical properties. This chapter reviews the intermetallic compounds formed during the manufacture of lead-free alloys. Intermetallic compounds formed in ordinary lead-based alloys are also discussed. The role of rare earth, especially indium and lanthanum, additions on intermetallic formation is examined. Microstructures of intermetallics are analysed. Hardness values of lead-free alloys are compared with emphasis on type and nature of intermetallics. SEM photographs of lead-free solders are discussed with regard to type of fracture, and the role of intermetallics in nature of fracture is examined. Lastly, the general mechanisms of formation of intermetallics are touched upon, and these mechanisms are extended to intermetallic formation in lead-free alloys.

Keywords: lanthanum based alloy, indium based alloy, hardness, characterization

1. Introduction

Intermetallic is a type of metallic compound that exhibits definite stoichiometric ratio of atoms. It also has a definite crystal structure. Their role is twofold. In small quantities, they can strengthen the soldering joint. However, when the amount of intermetallics increases, they may render the joint brittle. For example, La_3Sn intermetallics are known to be present in lanthanum-based lead-free alloys. These are known to increase the hardness of the alloys. Ag_3Sn, Cu_3Sn and Cu_6Sn are some known intermetallics in tin-, copper- and silver-based lead-free alloys. Mostly, the interaction of these intermetallics with the substrates has to be studied in detail in order to determine the exact effect on the solder properties. The growth of these intermetallics with time is determined by solid-state diffusion rules and varies from system to system.

2. Intermetallics in lead-based and lead-free solders

Way back in 1972, Lois Zackreysak has reported intermetallics in lead-free alloys. Tin-based lead-free alloys showed the presence of Sn—Cu intermetallic when soldered on a copper-based substrate. It was found that throughout the early stages of the growth process, the Cu_3Sn and the Cu_6Sn_5 thicknesses are approximately equal. As tin is depleted from the solder matrix, Cu_6Sn_5 growth slows down, while the Cu_3Sn continues to grow at the expense of the tin-rich intermetallic. The presence of insoluble alloying elements affects the intermetallic growth rate to a minor degree. The general theory for the diffusion-controlled growth of plates appeared to be applicable to this metallic system [1].

Just after soldering, Cu_6Sn_5 formed between the solder and copper pad. It should be noted that in most of the soldering joints, there are generally three layers formed. The three layers are substrate, intermetallic and solder. The intermetallic layers are sandwiched between the two substrate layers. In the study mentioned above, microwave energy has been used for soldering [2].

3. Growth kinetics

A study of growth kinetics of any solidification process will be useful to predict the type of microstructure formed and the morphology of the microstructures. Generally, as the alloys solidify under normal cooling conditions, we can expect a dendritic solidification morphology, just like in castings.

During soldering of lead-free alloys, the Cu_6Sn_5 intermetallic sublayer is clearly visible, and the Cu_3Sn sublayer is noticeable only for the sample annealed at 150°C [1].

Comparison of the 63Sn—37Pb joints and the lead-free joints shows that the initial thickness of the intermetallic layer is not significantly impacted by the higher temperature used during the lead-free assembly process. All the values are in a range of 1.6–2.3 microns [3].

The growth of these intermetallic layers can be modelled using parabolic growth kinetics [4]:

$$w = w_0 + D\sqrt{t} \tag{1}$$

where w = thickness of the intermetallic layer, w_0 = initial thickness of the layer, D = diffusion coefficient and t is the time of annealing.

Arrhenius type of growth kinetics is seen.

Li et al. have studied the growth kinetics of Sn-based lead-free solders on copper substrate. The results show that IMC layer is formed at solder-Cu substrate interface within a short time. It was found that Fick's law was not followed. Fick's law states that the mean total thickness increases linearly with the square root of the time. In most of the systems during solidification processing, welding and other manufacturing processes tend to follow Fick's law under equilibrium or near equilibrium conditions. In the case of soldering, cooling is very fast, especially when copper is used as substrate. It deviates from Fick's law at the early stage of the growth process and then approaches the parabolic law [5].

The growth behaviour of intermetallic compounds (IMCs) at the liquid-solid interfaces in Cu/Sn/Cu interconnects during reflow at temperatures in the range of 200–300°C on a hot plate, which was investigated by Zhao et al. The interfacial IMCs showed clearly asymmetrical growth during reflow. The growth of Cu_6Sn_5 IMC at the cold end was significantly enhanced while that of Cu_3Sn IMC was hindered especially at the hot end. It was found that the temperature gradient had caused the mass migration of Cu atoms from the hot end towards the cold end, resulting in sufficient Cu atomic flux for interfacial reaction at the cold end while inadequate Cu atomic flux at the hot end. The growth mechanism was considered as reaction/thermomigration-controlled at the cold end and grain boundary diffusion/thermomigration-controlled at the hot end [6].

Guo et al. [7] found that the interfacial Cu_6Sn_5 was much thicker at the cold end, whereas the consumption of Cu was much faster at the hot end in Cu/Sn—2.5Ag/

Cu solder joints during reflow at 260°C on a hot plate, due to the rapid migration of Cu atoms under a simulated temperature gradient of 51°C/cm. Qu et al. [1, 8] in situ studied the soldering interfacial reactions under a temperature gradient of 82.2°C/cm at 350°C using synchrotron radiation real-time imaging technology, and asymmetrical growth and morphology of interfacial IMCs between the cold and hot ends were clearly observed.

4. Hardness results

Hardness, especially microhardness is a very simple engineering measurement that gives information about the material being tested. In the case of volume of measurement being small as is the case of solders, then microhardness measurements are done from the periphery to the centre along any diagonal. In the case of solders, hardness values are an indication of whether intermetallic phases are present or not. For example, if the hardness of a material of a particular composition is higher than the normal solder alloy of the same composition, there is a possibility of intermetallic formation. Microstructural analysis should also be used to confirm the presence of intermetallics.

These hardness results are taken from the author's research work on lead-free alloys. All the values quoted here are experimental values obtained by the author and his team of coresearchers, and they have been published recently [9, 10].

4.1 Indium-based Sn—Zn alloys

All hardness values described in this section are taken along two perpendicular diagonals of a square-shaped specimen. Results are shown in (**Table 1**).

Sn—37Pb solders normally used for soldering have a hardness of 12 H_V [11]. So, on an average, both the new alloys have hardness higher than the traditionally used lead-based soldering alloys.

Hardness values are shown below in **Tables 2** and **3**.

Small changes in Zn and Al do not change the hardness much, and literature has shown that even small additions of lanthanum affect the hardness. So, it can be assumed with reasonable certainty that the hardness changes are due to changes in lanthanum content. Small additions of indium have been shown to improve the microhardness and also produce considerable changes in the microstructure [12]. Small amounts of indium addition have been found to refine grain size and improve hardness [13, 14].

Specimen	Hardness, H_V 0.2					Average
Sn—7Zn—3Al—3In	18.4	19.9	19.1	18.8	19.3	19.1
Sn—6Zn—2Al-2.5In	18.6	17.7	18.1	18.3	17.9	18.1

Table 1.
Results of hardness testing.

Specimen 88Zn—7Zn—2Al—2.5In	Hardness, H_V 0.2				
Diagonal 1	22.42	22.46	24.32	19.92	20.53
Diagonal 2	22.42	19.19	25.21	16.89	19.5

Table 2.
Hardness value of specimen 88Sn—7Zn—2Al—2.5In.

4.2 Lanthanum-based lead-free alloys

Two solders are analysed for their hardness. Both are lanthanum-based alloys with similar composition. Results are given in **Tables 4** and **5**.
The average hardness for sample 1 is **17.8 HV**.
The average hardness for sample 2 is **18.4 HV**.
Muhammed Aamir et al. [15] investigated that the inclusion of 0.4 wt.% of La in to Sn—Ag—Cu (SAC) solder system results in intermetallic growth. Intermetallics are hard in nature. Hence, our findings are in line with reported work, which indicates increase in hardness on addition of lanthanum.

4.3 Sn—Cu lead-free alloys

Referring **Table 6** above silver and manganese content are more in sample 1. Manganese is known to increase hardness. This could be the reason for higher hardness in sample 1. Since hardness of sample 1 is higher than sample 2, this could indicate the presence of more intermetallics in sample 1. With the addition of La, the microhardness of β-Sn and eutectic area was enhanced from 13.8 to 16.4 Hv and from 16.8 to 18.8 Hv, respectively [16]. Cu_6Sn_5, Ag_3Sn and $MnSn_2$ are present in dendritic Sn-rich solid solution (ß Sn). These intermetallics are found in both samples as seen in the microstructures of the samples. However, % age of intermetallics in each of the samples may be different. Average hardness of the second sample is considerably lower than the first one. This indicates that the presence of intermetallics is lesser in the second sample. More detailed analysis using XRD is required to

Specimen 88Sn—7.5Zn—2.5Al—2In		Hardness, H_V 0.2			
Diagonal 1	17.5	20.9	22.42	22.52	21.9
Diagonal 2	18.6	21.6	21.6	22.41	22.2

Table 3.
Hardness value of specimen 88Sn—7.5Zn—2.5Al—2In.

S. No	Diagonal 1 (μm)	Diagonal 2 (μm)	Microhardness value (HV 0.5) average
1	17.8	17.0	17.4
2	17.45	18.45	17.9
3	20.1	20.5	20.3
4	17.5	15.5	16.2
5	17.8	16.8	17.4

Table 4.
Microhardness value for sample 1 of composition 1 Sn-87.5% Zn-7.5% Bi-4% La-1%.

Diagonal 1 (μm)	Diagonal 2 (μm)	Microhardness value (HV 0.5)
16.5	16.5	16.5
17.3	18.1	17.7
18.6	22.6	20.2
19.5	19.0	19.3
17.6	19.6	18.2

Table 5.
Microhardness value for sample 2 of composition 2 Sn-87% Zn-7% Bi-4.5% La-1.5%.

S No	1	2	3	4	5	6	Avg
Sample 1	26.7	26.4	28.4	32.5	28.5	30.4	28.8
Sample 2	18.1	19.0	17.8	18.3	17.8	15.4	17.7

Table 6.
Hardness values for the two alloys.

Diagonal 1 (μm)	Diagonal 1 (μm)	Microhardness value (HV 0.5)
22.42	22.42	19.3
22.46	19.19	16.7
24.32	25.21	18.2
19.92	16.89	17.4
22.32	19.05	20.4

Table 7.
Microhardness value for sample 1.

Diagonal 1 (μm)	Diagonal 1 (μm)	Microhardness value (HV 0.5)
17.5	18.6	16.8
20.9	21.6	17.3
22.42	21.6	20.5
22.52	22.41	19.1
23.44	24.03	17.2

Table 8.
Microhardness value for sample 2.

come to a final conclusion. Zhang et al. showed that addition of 0.05% rare earth Yb suppressed the growth of intermetallics and the morphology of Cu_6Sn_5 layer can be changed to a relatively flat morphology (**Table 6**) [17].

4.4 Sn—Zn-lanthanum lead-free alloys

Two samples with slight compositional differences have been analysed, and the results were presented in **Tables** 7 and 8.
Sample 1 has a composition of **Sn-87.5% Zn-7.5% Bi-4% La-1%**.
The average hardness for sample 1 is **18.4 HV**.
Sample 2 has a composition of **Sn-87% Zn-7% Bi-4.5% La-1.5%**.
The average hardness for sample 2 is **18.2 HV**.
For Sn—8Zn—3Bi, with increasing temperature, the amount of hard Bi segregation increases which is the main cause of the rise in hardness.
Microstructral analysis done in a recent research work by the authors has been analysed.

5. Microstructural and EDAX analysis of Sn—Zn—Al-In alloys

5.1 Analysis of low Al-low indium alloys

Results of experimental work on 88Sn—7.5Zn—2.5Al—2In and 88Sn—7Zn—2Al—2.5 in lead-free soldering alloys are presented below.

Microstructures taken at different locations in the prepared solders showed variations, and the detailed analysis of the microstructures is given below.

Figure 1 shows the microstructure, and **Figure 2** shows the corresponding EDAX line spectrum. The weight % of sigma phase is 0.6%, when we consider the aluminium peak and 1.55% in the zinc peak. The indium cluster is seen in the middle and this showed 0.23% sigma phase. Sigma phase is a hard, brittle phase and generally not desirable. The presence of indium seems to have decreased the amount of sigma phase.

Indium peaks are not seen in **Figures 3** and **4**, when EDAX image is considered, and sigma phase is around 0.93% in both the Zn peak(first from the left) and the aluminium peak(second from the left). Sigma phase % age is more in this case compared to the previous microstructure (**Figures 5** and **6**).

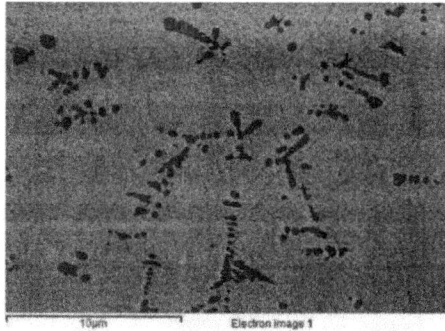

Figure 1.
Spectrum 1 electron image.

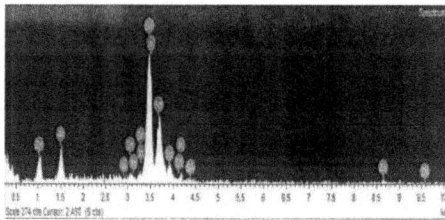

Figure 2.
Spectrum 1 EDAX graph.

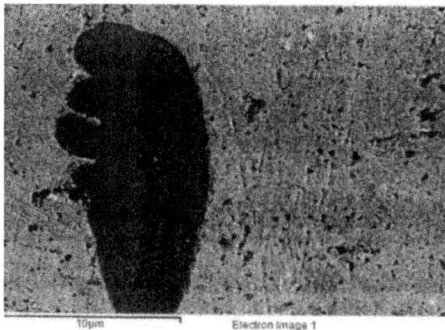

Figure 3.
Spectrum 2 electron image.

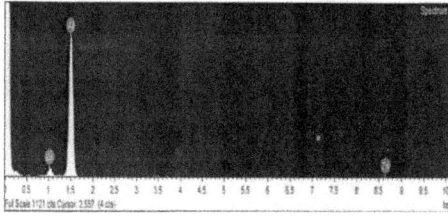

Figure 4.
Spectrum 2 EDAX graph.

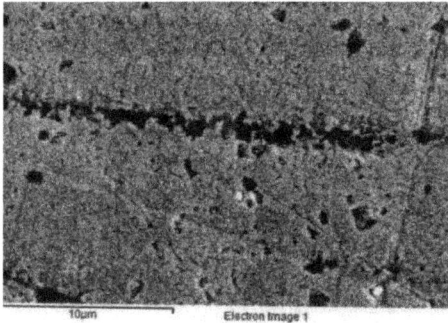

Figure 5.
Spectrum 3 electron image.

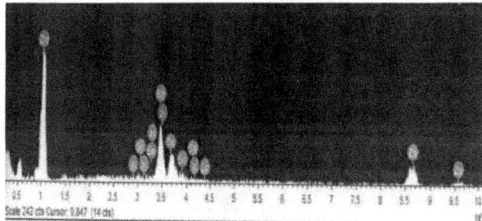

Figure 6.
Spectrum 3 EDAX graph.

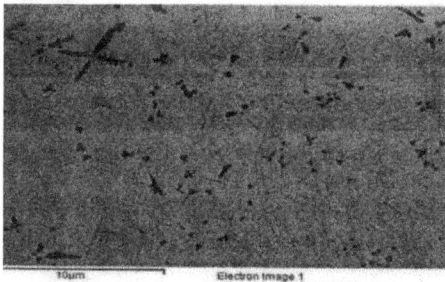

Figure 7.
Spectrum 10 electron image.

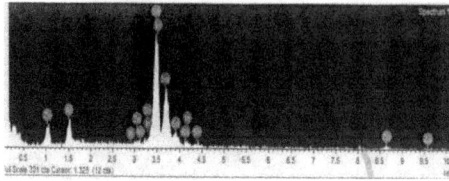

Figure 8.
Spectrum 10 EDAX graph.

It is seen that 1.91% sigma phase is present in the Zn peak, whereas the indium peak, which occurs as two small peaks in the tin cluster, shows 0.20% sigma phase.

Figures 7 and **8** show the microstructure and EDAX at another location. Here, sigma phase is 1.50% in the Zn peak and 0.55% in the aluminium peak. Indium peaks show 0.1% sigma and 1.425 sigma, respectively.

5.2 Analysis of high Al-high indium alloys

Similar results on the microstructure and EDAX analysis of 87Sn—7Zn—3Al—3In and 87.5Sn—6Zn—2Al—2.5 in lead-free soldering alloys are presented below.

5.2.1 Microstructural analysis

Needle shaped Zn phase and leaf shaped Aluminium phase are equally distributed in the microstructure. with the former contributing to increase in strength of the alloy. In the first alloy, the chemical composition of the needle- and leaf-shaped morphologies is given as follows: Leaf-shaped morphology (Al, 88%; Zn, 12%) and needle-shaped morphology (In, 72%; Zn, 27.96%; and Sn, 0.02%). In the second alloy, composition of the leaf- and needle-shaped morphologies is given as follows:

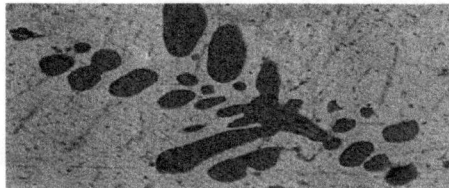

Figure 9.
Examination on leaf shape of the specimen.

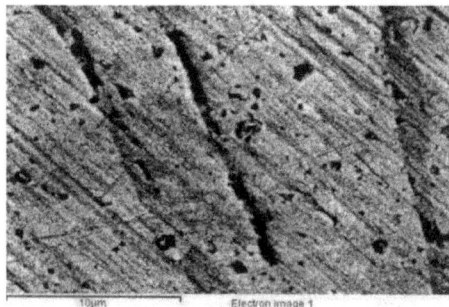

Figure 10.
Examination on needle shape of the specimen.

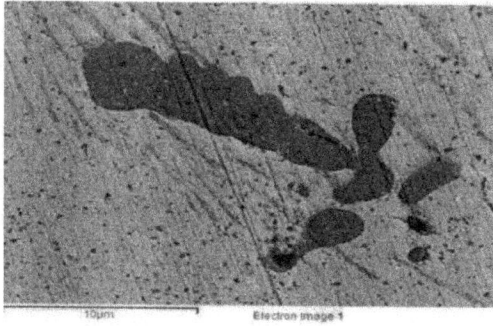

Figure 11.
Examination on leaf shape of the specimen.

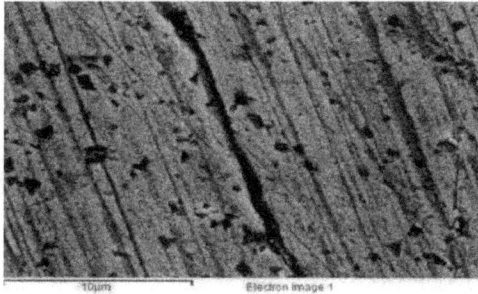

Figure 12.
Examination on needle shape of the specimen.

Leaf-shaped morphology (Al, 91.6%; Zn, 8.4%) and needle-shaped morphology (In, 66.36%; Zn, 31%; and Sn, 2.64%) (**Figures 9–12**).

5.2.2 EDAX analysis

EDAX analysis for both the alloys is presented in **Figures 13–16**. In both cases, the leaf-shaped morphology showed no intermetallics, whereas the needle-shaped morphology showed intermetallics.

Rod-like zinc-rich phases have been observed when cerium and lanthanum of the order of 0.1 wt.% are added [18]. In the current work, since the weight % of rare earth element indium added is relatively high, in all probability, the needle-like phase is rich in indium (72 and 66.36%, respectively) and zinc percentage is lesser,

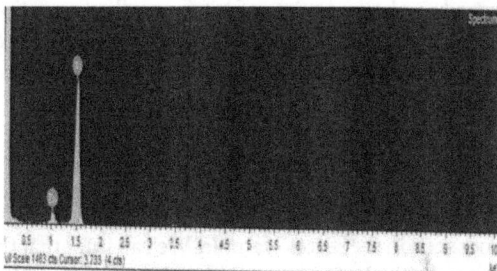

Figure 13.
Chemical composition of the leaf-shaped specimen-alloy 1.

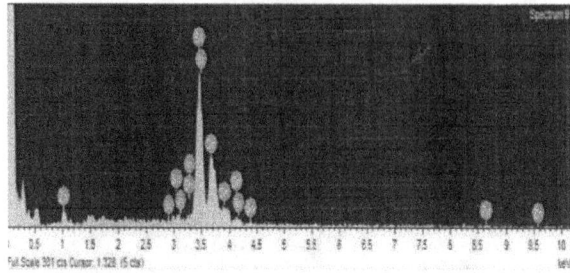

Figure 14.
Chemical composition of the needle-shaped specimen-alloy 1.

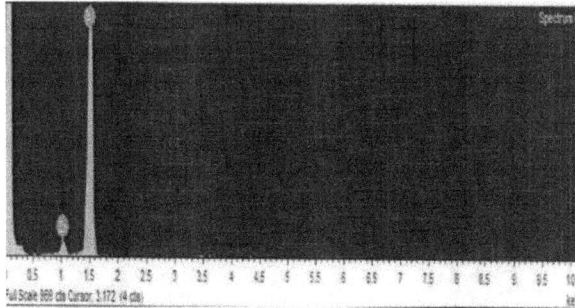

Figure 15.
Chemical composition of the leaf-shaped specimen-alloy 2.

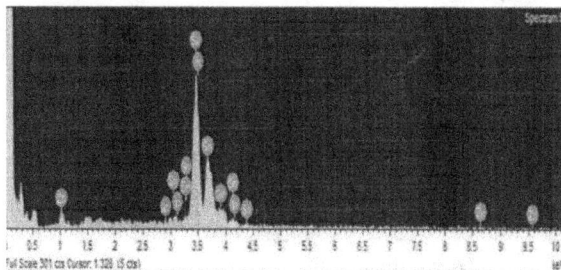

Figure 16.
Chemical composition of the needle-shaped specimen-alloy 2.

agreeing with published literature on cerium and lanthanum addition. Gadolinium addition has been found to refine grains of Mg—5Sn—Zn—Al [19].

In some cases, there have been reports of intermetallic formation suppression as a result of the addition of rare earth metals. A study by Huan Lee et al. showed that addition of praseodymium reduced the formation of intermetallics at the junctions of a Sn—Zn—Ga solder [20].

5.3 Microstructural analysis of Sn—Cu lead-free solders

Figure 17(a, b) shows the dendritic structure of 87Sn—7Cu—3.0Mn—3Ag. The dendrite structure is not seen in the second sample. We can infer that the dendrites are broken down in the second sample. Han et al. [10] and Yang et al. [21] also have reported the influence of Ni-coated carbon addition on the microstructure of Sn3.5Ag0.7Cu nanocomposite solder. It is found that the morphology of Ag_3Sn and Cu_6Sn_5 was

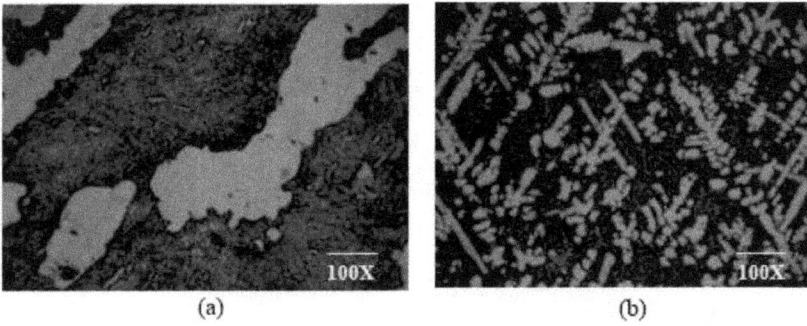

Figure 17.
Microstructures of (a) 87Sn—7Cu—3.0 Mn ig. 5. Microstructures of (b) 87.5Sn—7.5Cu—2.5Mn—2.5Ag.

uniformly distributed in the solder matrix. But the addition of another fourth element to form a quaternary changes the microstructure. Some researchers produced intermetallics, which react with Sn, thus refining the microstructure of Sn, Ag and Cu solder. Other researchers chose to add some low solubility and diffusivity in Sn, such as Al_2O_3, TiO_2, SiC and POSS. In general, it was found that until a critical value of alloying element content, properties are enhanced, and above this % age, it becomes detrimental to the alloy.

So, addition of Manganese seems to have increased the melting point. The slight difference in melting temperatures could be due to the difference in alloying elements in each sample. Addition of 0.2% iron to SnAg Cu alloys results in two exothermic peaks at 220 and 235°C. When adding 0.6 wt.% Fe, only a single endothermic peak at 221.35°C was found, showing that it has a eutectic composition [22]. Based on the ternary SnAgCu phase diagram [23], there are two steps in the DSC. This could be because of melting of ternary eutectic β-Sn + Ag_3Sn + ηCu_6Sn_5 phase and melting of primary β-Sn.

6. SEM fractography

SEM micrographs are generally used to give information about the type of fracture. Here, since tensile tests have not been done, SEM has been used to identify Sn whiskers. There have been some studies showing Sn whiskers in the SEM micrographs of SnCu alloys.

The abovementioned SEM micrographs are of two samples of slightly varying chemical composition. **Figures 18** and **19** show the absence of tin whiskers in both

Figure 18.
Sample 1.

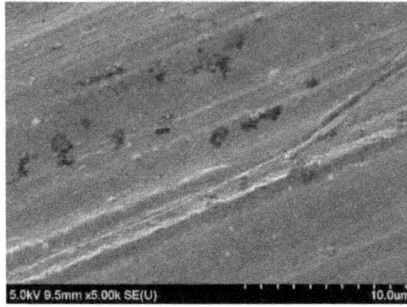

Figure 19.
Sample 2.

the alloys. The presence of tin whiskers results in deterioration of electrical properties. Chuang and Lin [24] showed that 0.5% Zn addition into SnAgCu solder refined the microstructure and suppressed whisker growth.

Sn—Zn alloys have been used as a substitute for Pb—Sn alloys. However, Sn—Zn alloys are susceptible to oxidation and ZnO forms easily (**Figures 20** and **21**) [25–27].

SEM micrographs are for two different compositions of Sn—Zn—La lead-free solders (composition as mentioned in the Hardness **Tables 4** and **5**). Some

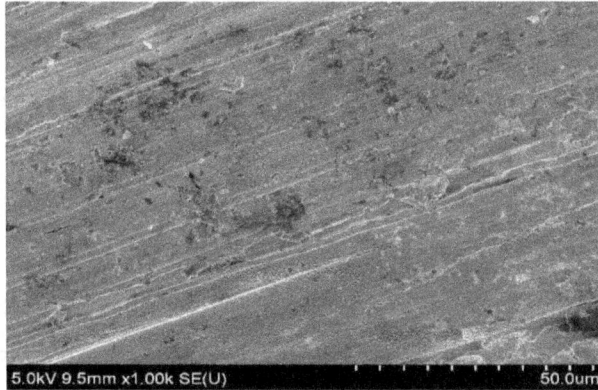

Figure 20.
SEM micrographs of Sn—Zn—La alloys.

Figure 21.
SEM micrographs of Sn—Zn—La alloys.

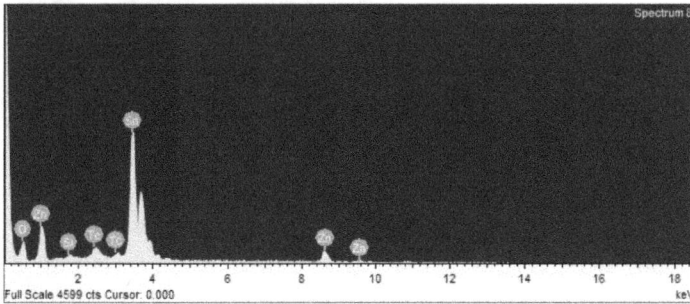

Figure 22.
EDAX of Sn—Zn—La sample.

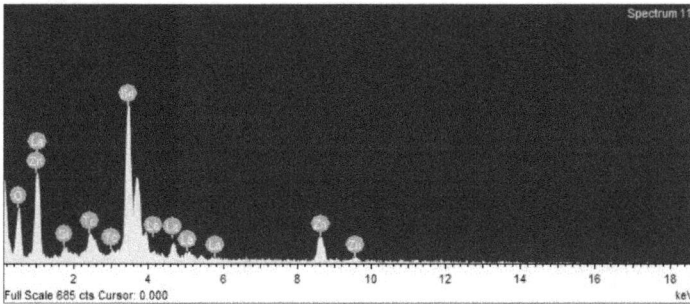

Figure 23.
EDAX of Sn—Zn—La of slightly different composition.

precipitates are seen, which may have an effect on the hardness and tensile strength of the solders. However, in this case, it was found that hardness for both alloys is around 18 H_V. Hence, the effect of precipitates seems to be marginal for this lead-free alloy. These SEM micrographs have been taken from unpublished work done by the author as part of undergraduate project work.

EDAX Studies.

EDAX Studies of two samples (**Figures 22** and **23**).

There is not much difference between the EDAX of the two samples. There do not seem to be intermetallics present in this alloy. Hardness values also corroborate this statement.

7. Conclusion

This chapter starts with the importance of intermetallic systems in lead-free solders. The effect of some rare earths like lanthanum and indium on intermetallic formation in lead-free alloys has been explained. Tin-zinc-lanthanum and tin-zinc-aluminium-indium have been chosen for detailed analysis. Microhardness surveys have been done along two perpendicular diagonals of a square-shaped specimen, and the results of the microhardness surveys have been analysed. Microstructures in both these systems have been discussed with respect to the phases present and their distribution. Based upon both the microhardness and microstructure of the two alloys chosen, the presence of intermetallics and the types of intermetallics has been discussed. Growth kinetics of intermetallics has been briefly analysed. SEM fracto-graphs have been taken to show the type of fracture. EDAX has been employed to show the distribution of phases.

Author details

Dattaguru Ananthapadmanaban[1]* and Arun Vasantha Geethan[2]

1 Department of Mechanical Engineering, SSN College of Engineering, Chennai, India

2 Department of Mechanical Engineering, St. Josephs Institute of Technology, Chennai, India

*Address all correspondence to: ananthapadmanaban.dattaguru@gmail.com

IntechOpen

References

[1] Zakraysek L. Intermetallic Growth in Tin-Rich Solders, Welding Research Supplement; 1972. pp. 536-541

[2] Faizal SM, Azlina OS. Intermetallic compound formation on lead-free solders by using microwave energy. AIP Conference Proceedings. 2016;**1774**:060008

[3] Roubaud P, Ng G, Bulwith GH-HPR, Prasad RH-AMS, Kamath FC-CPACS, Garcia-Sanmina A. Impact of intermetallic growth on the mechanical strength of Pb-free BGA assemblies. In: APEX 2001 International Conference. San Diego; 2001

[4] Wassink R, Klein J. Soldering in Electronics. 2nd ed. Ayr, Scotland: Electrochemical Publications Limited; 1989

[5] Li GY, Chen BL. Formation and growth kinetics of interfacial intermetallics in Pb-free solder joint. IEEE Transactions on Components and Packaging Technologies. 2003;**26**(3):651-658

[6] Zhao N, Zhong Y, Huang ML, Ma HT, Dong W. Growth kinetics of Cu_6Sn_5 intermetallic compound at liquid-solid interfaces in Cu/Sn/Cu interconnects under temperature gradient. Scientific Reports. 2015;5:13491

[7] Guo MY, Lin CK, Chen C, Tu KN. Asymmetrical growth of Cu_6Sn_5 intermetallic compounds due to rapid thermomigration of Cu in molten SnAg solder joints. Intermetallics. 2012;**29**:155-158

[8] Qu L, Zhao N, Ma HT, Zhao HJ, Huang ML. In situ study on the effect of thermomigration on intermetallic compounds growth in liquid-solid interfacial reaction. Journal of Applied Physics. 2014;**115**:204907

[9] Arthur Jebastine Sunderraj D, Ananthapadmanaban D, Arun Vasantha Geethan K. Chapter 30: Comparative hardness studies and microstructural characterization of 87 Sn-7Zn-3Al-3In and 87.5 Sn-6 Zn-2Al-2.5 in lead free soldering alloys. In: Advances in Materials and Metallurgy Lecture Series in Mechanical Engineering. Springer Nature. pp. 311-323

[10] Arthur Jebastine Sunderraj D, Ananthapadmanaban D, Arun Vasantha Geethan K. Preparation, hardness studies and characterization of 88 Sn-7.5Zn-2.5Al-2In and 88 Sn-7 Zn-2Al-2.5 in lead free soldering alloys. In: IOP Conference Series, Materials Science and Engineering. 2018

[11] Steward T, Liu S. Database for Solder Properties with Emphasis on New Lead Free Solders. National Institute for Standards and Technology and Colorado School of Mines; 2002

[12] Budi Dharma IGB, Hamdi M, Ariga T. The effects of adding silver and indium to lead-free solders. Welding Journal. 2009;**88**(4):45-47

[13] Shalaby RM. Effect of indium content and rapid solidification on microhardness and micro-creep of Sn-Zn eutectic lead free solder alloy. Crystal Research and Technology. 2010;**45**(4):427-432

[14] Wang Y-T, Ho CJ, Tsai H-L. Effect of in addition on mechanical properties of Sn-9Zn-In/Cu solder. In: 8th IEEE International Conference on Nano/Micro Engineered and Molecular Systems. Nano/Micro Engineered and Molecular Systems (NEMS); 2013

[15] Aamir M, Muhammed R, Ahmed N. Mechanical properties of lead free solder alloy for green electronics under high strain rate and thermal aging. Journal of Engineering and Applied Science. 2017;**36**(1):115-123

[16] Zhang L, Fan XY, Guo YH, He CW. Properties enhancement of SnAgCu solders containing rare earth Yb. Materials and Design. 2014;**57**:646-651

[17] Tay SL, Haseeb ASMA, Johan MR, Munroe PR, Quadir MZ. Influence of Ni nanoparticle on the morphology and growth of interfacial intermetallic compounds between Sn-3.8Ag-0.7Cu lead-free solder and copper substrate. Intermetallics. 2013;**33**:8-15

[18] Wu CML, Law CMT, Yu DQ, Wang L. The wettability and microstructure of Sn-Zn-RE alloys. Journal of Electronic Materials. 2003;**32**(2):63-69

[19] Fang CF, Meng LG, Wu YF, Wang LH, Zhang XG. Effect of Gd on the microstructure and mechanical properties of Mg-Sn-Zn-Al alloy. Applied Mechanics and Materials. 2013;**312**:411-414

[20] Ye H, Xue S, Luo J, Lee Y. Properties and interfacial microstructure of Sn-Zn-Ga solder joint with rare earth Pr addition. Materials and Design. 2013;**46**:316-322

[21] Yang ZB, Zhou W, Wu P. Effects of Ni-coated carbon nanotubes addition on the microstructure and mechanical properties of Sn-Ag-Cu solder alloys. Materials Science and Engineering A. 2014;**590**:295-300

[22] El-Daly AA, Hammad AE, Al-Ganainy GS, Ragab M. Influence of Zn addition on the microstructure, melt properties and creep behavior of low Ag-content Sn-Ag-Cu lead free solders. Materials Science and Engineering A. 2014;**608**:130-138

[23] Shnawah DA, Sabri MFM, Badruddin IA. A review on thermal cycling and drop impact reliability of SAC solder joint in portable electronic products. Microelectronics Reliability;**52**(1):90-99

[24] Chuangand TH, Lin HJ. Inhibition of whisker growth on the surface of Sn-3Ag-0.5Cu 0.5Ce solder alloyed with Zn. Journal of Electronic Materials. 2009;**38**(3):420-424

[25] Hirose A, Yanagawa H, Ide E, Kobayashi KF. Joint strength and interfacial microstructure between Sn–Ag–Cu and Sn–Zn–Bi solders and Cu substrate. Science and Technology of Advanced Materials. 2004;**5**:267

[26] Iwasaki T, Kim JH, Mizuhashi S, Satoh M. Encapsulation of lead-free Sn/Zn/Bi solder alloy particles by coating with wax powder for improving oxidation resistance. Journal of Electronic Materials. 2005;**34**:647

[27] McCormack M, Jin S, Chen HS. Wetting interaction of Pb-free Sn-Zn-Al solders on metal plated substrate. Journal of Electronic Materials. 1994;**23**:687

Chapter 3

Soldering by the Active Lead-Free Tin and Bismuth-Based Solders

Roman Koleňák, Martin Provazník and Igor Kostolný

Abstract

The chapter deals with Sn and Bi-In-based lead-free solders. The term "active solders" is used for the solders which contain one or more elements with enhanced affinity to some element contained in the substrate material. Mainly, Ti, In, lanthanides, etc. belong amongst the active metals. The role of an active element is to ensure a good wetting by a reactive decomposition of the surface layer of substrate. The perspective solders for joining the combined materials, as ceramics/metal, are mainly the tin-based, lead-free solders, which are enriched with titanium (usually up to 4 wt. %). The advantage consists in the fact that they offers a sufficient plasticity reserve, by what they are capable to compensate undesired residual stresses formed in the joint. Titanium also reacts with carbon, nitrogen or oxygen of the ceramic material, eventually it forms the intermetallic phases, which increase the strength of joint interface. The Sn-Ti, Sn-Ag-Ti and Bi-In-Sn solders were selected for the experiments. These solders were applied for fabrication of Al_2O_3 ceramics/Cu joints. The phase composition and microstructure of solders and soldered joints was analysed. Interactions in the interface of ceramic/solder and Cu substrate/solder were determined. The shear strength of soldered joints was measured

Keywords: active solder, ceramics, metals, ultrasonic soldering, shear strength

1. Introduction

1.1 Soldering with active solders

The term 'active solders' is occurring in the technologies oriented to fabrication of combined joints already for several decades. These solders contain an active element which reacts with the surface of the parent material during soldering process. This reaction takes place owing to the fact that an active element exerts higher affinity to the elements in the chemical composition of the substrate. The role of an active element is to ensure a good wetting of substrate with solder by reactive decomposition of the surface layer of the parent metal and by reducing the interfacial stress in the ceramics-solder interface. The active solder then may be used for fabrication of joints with different, either ceramic or metal, substrates. The most used active metals are titanium, zirconium or hafnium [1]. The chemical bonds in the interface of an active solder (with titanium content) and a solid substrate (ceramics/metal) are shown in **Figure 1**. Concentration of an active element should be sufficiently high in order to cause the wetting on a ceramic substrate, while it must not cause the formation of brittle intermetallic

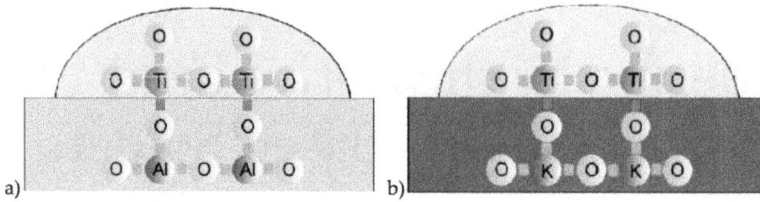

Figure 1.
Chemical bonds in the interface of (a) solder-ceramics, (b) solder-metal.

phases. Though a good wetting is achieved at higher content of active elements in a soldering alloy, nevertheless the joint then exerts poorer mechanical properties [2]. The commercially used active solders usually contain relatively small amounts of active elements (not more than 4 wt. %). Addition of indium to solder supports the wetting of substrate and lowers the soldering temperature and thus also the thermal expansivity of the fabricated joint [3]. The additional elements with higher affinity to oxygen (cerium and lanthanum) may protect the active solders against excessive oxidation of an active element during soldering in the air. At the same time also, the elements supporting wetting, for example, gallium [4, 5], are added to active solders.

The active element is moved from the entire solder volume to both soldered materials during the soldering process. Thus, a reaction layer in the thickness of several μm, containing the reaction products of an active element and substrate, is formed in the interface of soldered joint. The thickness of this reaction layer depends on the solder type and soldering conditions. The active element with a high affinity to oxygen, which reacts with the ceramics during soldering, creates the bonds in an interatomic level. The active element (e.g. Ti) contained in the solder bonds the oxygen from the surface layers of oxide ceramics. The following oxide types are formed on the ceramics by the reaction of an active element: TiO, Ti_2O_3, Ti_3O_5, Ti_4O_5 and TiO_2. The reaction product alters the surface energy of ceramics and enables wetting of the solder [6]. The following type of reaction takes place between the active element and Al_2O_3 ceramics [7].

$$M/AO_x \rightarrow M/m_aMA_a/n_bM_bA_cO_d/AO_x$$

where
M   metal element of the solder/interface.
AO_x ceramic element
O   may be an oxide
n   different combined oxides
M   different intermetallic compounds

This type of chemical reaction is valid for the active soldering with an active element as Ti, Zr and/or Hf. The layer formed after such a reaction will depend on the soldering parameters (temperature and time) and environment atmosphere (vacuum oven and inert gas). For the active element Ti on Al_2O_3 substrate, it is valid [7]:

$$Ti/Al_2O_3 \rightarrow Ti/Ti_3Al/TiAlO_x/Al_2O_3$$

In the selection of an active element for soldering ceramic materials, it is necessary to take into account the fact that higher concentration of an active metal may result in joint embrittlement. This is caused by the formation of brittleness phases in the ceramic-solder zone [8].

The use of Ti as an active element in a solder has been investigated by many authors. Titanium in the $SnAgTi_4$ solder provided a reduction in the wetting angle to the sapphire substrate. The absorption of Ti together with the release of Al from the sapphire substrate provides an interaction between the solder and the substrate [9]. Another investigated aspect of the Sn-Ag-Ti-based solder was its spreadability on porous graphite when applying ultrasonic vibrations. The authors' results [10, 11] have shown that the application of ultrasonic waves during soldering allows the active solder to spread on the surface of the graphite at a relatively low temperature under atmospheric conditions. The oxide layer on the molten solder forms a compact layer of a certain thickness which prevents the liquid solder from wetting the base material. The reduction of Ti oxide formation is possible by the addition of Y, as documented by the authors [12]. They found that the oxide layer of the solder consists mainly of titanium dioxide. A small amount of Y improves the resistance to oxidation because it suppresses the oxidation of Ti in the molten solder and reduces the amount of oxygen atoms entering the molten solder. The composition of the surface oxide of $SnAg_4Ti_2Y_{0.5}$ solder was mainly Y_2O_3 and a low amount of titanium dioxide. After ultrasonic application, the oxide layer was disrupted, and the solder was able to wet the materials. The authors found that the spreadability and wettability of Sn-Ag-Ti-based solders on the graphite substrate could be improved by increasing the time of ultrasound activation. Similar observations were published in [13], where the authors deal with the soldering of aluminium-graphite composite material using a $SnAg_{3.5}Ti_4Cu_{0.5}$ solder at soldering temperature of 250°C. Aluminium from the composite was dissolved in the active solder and formed a solid solution of Al-Ag-Sn at the interface. The average shear strength of the Al-Gr/Al-Gr joints was 8.15 MPa.

The main factors affecting the solder selection comprise different melting points of soldered materials, different surface stresses of the substrates and the residual stresses formed during solidification process. Therefore such a solder is proposed where the matrix exerts a sufficient plasticity reserve, capable to compensate the residual stresses formed in the joint [14]. To overcome the mentioned problems, the solder meeting a complex set of desired criteria for the quality of resultant joint is selected. The active solders, similarly as the commercial solders, are classified by the melting temperature to solders, brazing alloys and high-temperature solders. The active solders are further classified by the mechanism of bond formation to high-temperature and mechanically activated ones. For reaction capacity of an active element, the soldering temperature in the case of high-temperature-activated solders must be higher than 780°C. The classification of active solders and their chemical composition is shown in **Table 1**.

1.2 Active solders

The primary materials of an active solder are usually tin, lead, bismuth, zinc or indium and the alloys based on these metals. Active solders allow to join also unusual combinations of metallic materials (e.g. CrNi steel, Mo, W, Ti, Cr, etc.) and non-metallic materials as siliceous glass, sapphire, carbon, silicon and several types of ceramics. In soldering with active solders, the solder is capable to compensate the stresses formed due to different thermal expansivity of joined materials by its plastic strain, shear mechanism or yield. In such a manner, the highest reduction of residual stresses may be achieved at a preserved joint stability. In the case of soldering untraditional material combinations with extremely different coefficients of thermal expansivity (e.g. glass with aluminium/copper), heavier solder thickness should be selected, in order to prevent the cracks in the joint interface. Such joints are used mainly in electrotechnics, where lower strength and thermal resistance of joint are sufficient. These solders also allow to fabricate the

Active solders	Application temperature	Chemical composition
Solders	From cryogenic temperatures up to 200°C	High-temperature activation, based on Sn, In and Pb (e.g. $Sn_{90}Ag_{10}Ti_3$)
		Mechanical activation based on Sn (e.g. $SnAg_{3.5}Ti_4(Ce,Ga)$)
Brazing alloys	Up to 400°C	Based on Ag, Cu and Au (e.g. $Ag_{72}CuTi_{1.5}$)
High-temperature solders	Up to 900°C	Based on Ni, Co, Pd and Pt (e.g. $Ni_{70}Hf_{30}$)

Table 1.
Classification of active solders by the melting point.

vacuum-tight joints used in the vacuum and cryogenic technology, where indium-based solder has proved as suitable. The solders containing lead are not suitable for soldering in vacuum, since considerable evaporation of lead and also oven contamination may occur. In the case of soldering with lead, it is necessary to employ the through-flow atmosphere of pure argon, eventually helium [15]. The active solders may be activated mechanically (by scrapping or ultrasound) in dependence on material which they wet (metal/ceramics) within the temperature interval from 200 to 400°C. An essential group of these solders applicable at lower soldering temperatures comprises the tin-based, lead-free solders which are enriched by a small amount of an active element as titanium. An example of such active solders destined for soldering a wide scope of materials is the solders type S-Bond, where also $SnAg_{3.5}Ti_4(Ce,Ga)$ solder [16, 17] may belong.

Titanium is moved from the solder matrix to the interface; combines with carbon, nitrogen or oxygen from the ceramic material; and thus creates intermetallic compounds which allow the wetting of ceramic substrate and creation of a metallurgical bond. Fluxless soldering process in the air is concerned, where no corrosion owing to flux remnants occurs [18, 19].

The active solders for high-temperature activation necessitate considerably higher soldering temperature and may be used only in a vacuum and/or shielding atmosphere. Soldering temperature of the tin- and lead-based solders varies within 850–950°C. The solders for mechanical activation (e.g. $SnAg_{3.5}Ti_4(Ce,Ga)$) may be used also for high-temperature activation. However, greater wetting angles than in the case of mechanical activation are attained [14]. As an example of using S-Bond 220-1 ($SnAg_{3.5}Ti_4(Ce,Ga)$) solder in electrotechnics, joining of Al_2O_3 with copper in **Figure 2** may be mentioned. This solder allows to join the materials with different

Figure 2.
Detailed view of Al_2O_3-Cu joint fabricated with S-Bond solder [20].

coefficients of thermal conductivity and expansivity (metals, light metals, ceramics, composites with metal matrix, carbides, glass, etc.) [20].

The solder type S-Bond contains Ti as an active element and elements from the group of lanthanides. These active elements migrate to the interface of the soldered joint and act upon the material surface by the active disruption and removal of oxides. Disruption of oxide layer, which prevents the formation of contact between the solder and substrate surface, is called as 'activation'. As soon as the oxide layer is disrupted, the solder volume reacts with the substrate surface, and a strong bond with the joined surface is thus formed.

The bond is formed by two mechanisms:

- *Metallurgic bonding*—for example, with Cu and/or Al surfaces. Besides the active element, also other elements of solder, as Sn-Ag-Ti, react with the substrate elements. These contribute to bond formation by creation of phases as Al-Ti, Cu-Sn, Ag-Sn, etc. system. This process may be used for a wide scope of metallic materials.

- *Gravity bond*—in the case of metals with very thin dielectric surface oxides. These are, for example, Ti or stainless steel with TiO_2 eventually Cr_3O_2 oxides. Bonding is ensured by the gravity of surfaces with opposite electric charge. The active elements of solder and the elements of joined substrate are attracted across the interface by the van der Waals forces.

1.2.1 High-temperature activation of an active solder

In the case of active solders containing Ti (or other reactive elements), the high-temperature activation in shielding atmosphere of vacuum is employed. The active element reacts with the surface oxides of substrates at the soldering temperatures of 850–950°C. Similarly also nitrides, carbides or silicides of the active element are formed in the solder in the case of non-oxide ceramics. One of the greatest disadvantages of this process consists in the necessity of a vacuum.

The work cycle in **Figure 3** consists of a rapid heating to soldering temperature, dwell time for about 9 minutes and a long free cooling down (for about 320 minutes) in the oven. The short dwell time at the soldering temperature prevents formation of brittle phases and grain coarsening. The slow cooling down is needed in order to prevent high residual stresses, which may cause the joint cracking.

The diffusion processes take place during slow cooling down, which will be exerted in the growth of diffusion zone in the parent metal and the solubility zone in the solder. Soldering with an active solder may be performed in a vacuum furnace at

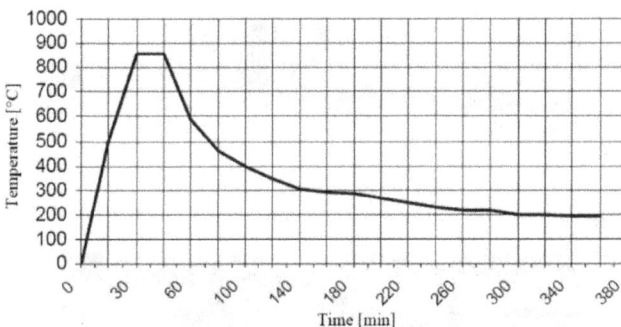

Figure 3.
Scheme of thermal cycle of vacuum soldering [21].

the pressure of 10^{-1}–10^{-3} Pa. In some special cases, soldering may be performed also in the overpressure of argon or helium. Nitrogen cannot be used as a shielding gas since it deteriorates the wetting of ceramics. This is caused by the fact that nitrogen has a high affinity to Ti, and thus it depletes the solder by Ti. The high-temperature activation cannot be used for soldering of metallic materials with the coatings of Au, Ag, Cu, Al and Mg, which exert a high dilution rate in Sn solder. The working parameters exert a significant effect upon the properties of soldered joints [21].

1.2.2 Mechanical activation of solder

Mechanical activation seems to be a new trend in the field of active solders at present. Soldering is realised at temperatures of 250–280°C with the dwell time from 30 seconds up to 3 minutes. The time- and power-demanding high-temperature activation is in this process replaced with the mechanical activation of an active element. In this way, the necessity of a vacuum, shielding atmosphere or multistep solder deposition is eliminated [21].

Mechanical activation may be realised by:

- scratching

- spreading with a metal brush

- vibrations

- ultrasound

The primary reason for activation consists in the fact that the surface of metallic materials is covered by an oxide layer, which must be gradually disrupted during the soldering process. Activation of surface layers on ceramic materials is possible exclusively by the application of ultrasound.

In order to allow soldering in the air without the necessity of flux, the active solder is alloyed with the elements from the group of lanthanides. These are the rare earth metals, for example, Ce and Ga, which protect Ti against oxidation during heating and soldering [22]. The soldered joint fabricated with $Sn_{35}Ag_4Ti(Ce,Ga)$ solder is shown in **Figure 4**.

The work cycle of soldering with ultrasound activation is considerably shorter than in the case of high-temperature activation. The soldering temperatures are also significantly lower than at high-temperature activation. The structure of soldered materials is less affected, and therefore lower residual stresses are formed.

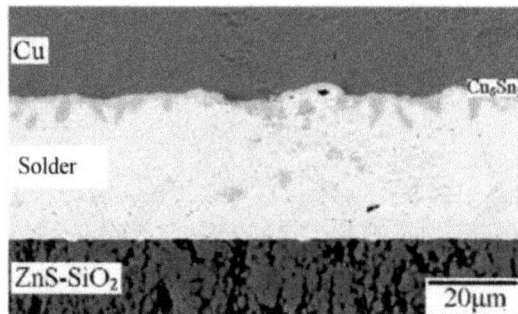

Figure 4.
Interface of ZnSiO$_2$, Sn$_{35}$Ag$_4$Ti(Ce,Ga), Cu joint [23].

1.3 Activation mechanism of an active element by ultrasound

The surface of soldered materials is covered with an oxide layer which must be gradually disrupted during soldering. This is performed by the mechanism called 'solder activation'. For the soldering of metals, it is mostly sufficient to activate mechanically by scratching; however, in the case of ceramic materials, it is necessary to employ the ultrasonic activation with the frequency over 20 kHz.

Mechanical activation is realised by:

- scratching and spreading with a metal brush (suitable for soldering metals as Cu, Al, Ni, CrNi steel, etc.)

- vibrations (50–60 Hz)

- ultrasound with the frequency over 20 kHz (suitable for soldering ceramic and non-metallic materials)

The most used technologies for fabrication of combined soldered joints type ceramics-metal are derived from ultrasonic soldering. This results from the finding that only ultrasonic activation is sufficiently efficient for disrupting the surface layers on ceramic materials [15, 21]. The physical principle of ultrasonic soldering consists in the fact that cavitation of sufficient intensity occurs in liquids and molten metals affected by ultrasonic field. The erosive activity of cavitation attacks, disrupts and removes the oxides from the surface of the soldered part. If a solder with a sufficient content of active elements is used, the reliable, diffusion and metallurgical bonding with the parent material is attained. Ultrasonic cavitation reduces the surface tension and enhances the spreadability and capillarity of solders. It also significantly affects the distribution of an active element in the solder matrix and supports the diffusion processes in the phase interface. The time of solder activation by ultrasound partially depends on the resistance of surface oxide layers against the cavitation erosion. However, the times of working cycles are incomparably shorter than the times of activation at high temperatures in vacuum. Application of ultrasonic method is sometimes limited by the soldering material used. In the case of brittle substrates, the damage of specimen by cracking the surface layers may occur [24].

1.3.1 Principle of solder activation by power ultrasound

Majority of power ultrasound applications, where also ultrasonic soldering belongs, necessitate semi-wave transducers with the resonance frequencies of 20–60 kHz. Ultrasonic transducer transfers the electric power to mechanical—the so-called ultrasonic—oscillations. Ultrasonic head consists of an oscillating system fastened in a case made of plastic. The protective case serves for an ergonomic grasping of the tool, eventually its clamping on a stand. The oscillating system is formed by a piezoelectric transducer, a concentration adapter and an exchangeable working tool. The exchangeable tools—sonotrodes—which are screwed on the adapter may be of different shapes. In most cases it is a conical point made of titanium alloy [25]. The principal scheme of an equipment for ultrasonic soldering is shown in **Figure 5**.

The sonotrode point is oscillating with the frequency of alternating current supplied by the generator through the connecting cable. The amplitude of oscillation is variable, and it is altered with the frequency of the supplied current. It generally varies at the level of 10^{-4} mm. The rate and intensity of applied UT oscillation vary within a certain range, and it is selected with regard to the process conditions and the character of materials. In this way, it is possible to affect both the strength characteristics of the

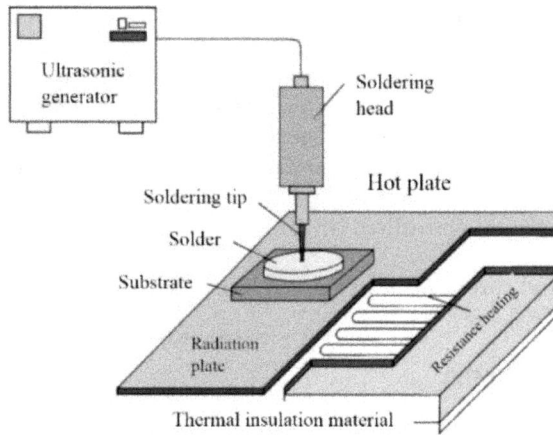

Figure 5.
Principal scheme of an equipment for ultrasonic soldering.

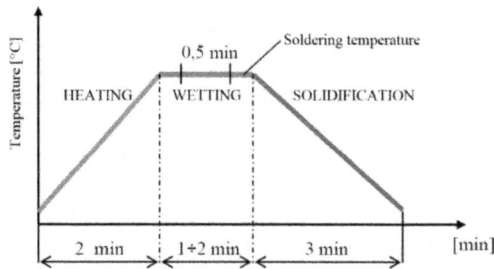

Figure 6.
Work cycle of soldering with application of mechanical activation.

joint and its technological parameters (e.g. the joint width) as well [26]. The work cycle of soldering with application of mechanical activation by ultrasound is roughly 10 times shorter than in the case of high-temperature activation (**Figure 6**).

Acting upon the active solder by ultrasound may be performed in several ways. The technology depends on the size and geometry of the joint, soldering temperature, type of parent metal, number of produced pieces and other parameters. The most used technique of laboratory activation of melts by ultrasound comprises the method, where the solder is molten on the surface of a joined unit by an external heat source, and the ultrasonic power is supplied through the solder from a sonotrode point on the part surface. This method makes it possible to solder selectively the localised surfaces on large parts or to cover the entire surfaces of parts. This principle is shown in **Figure 7** [24].

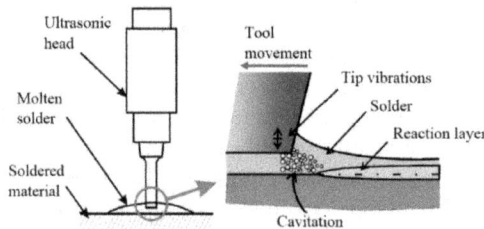

Figure 7.
Manual soldering with ultrasound assistance.

2. Active solders for ultrasonic activation

2.1 SnTi$_2$ solder

Figure 8 shows the microstructure of SnTi$_2$ solder. This solder was manufactured by free casting in a vacuum. The solder exerts heterogeneous composition. The matrix consists of 100% Sn. The round, dark and oblong grey phases are the intermetallic compounds of the Ti-Sn system. The zones of solder containing 35.2% Ti and 64.8% Sn consist of Ti$_6$Sn$_5$ phase. Composition of SnTi$_2$ solder in the selected spots is shown in **Table 2**. Diffraction analysis performed on a sample of SnTi$_2$ (**Figure 9**) solder has revealed the presence of the following phases: Ti$_6$Sn$_5$, Ti$_3$Sn, Sn$_3$Ti$_5$ and Sn$_5$Ti$_6$. Solder microstructure consists of a tin matrix (100% Sn) with non-uniformly distributed phases of the Ti-Sn system. The coarse dark-grey phases in **Figure 8** represent the Ti$_6$Sn$_5$ phase; the smaller bright-grey phases with Sn designation contain 3.6% Ti and 96.4% Sn.

2.2 SnAg$_{3.5}$Ti$_4$(Ce,Ga) solder

The SnAg$_{3.5}$Ti$_4$(Ce,Ga) solder shown in **Figures 10** and **11** consists of a tin matrix with non-uniformly distributed constituents of intermetallic phases of binary Ti-Sn, Ag-Sn and Ag-Ti systems. The dark-grey phases contain an average of 31.5% Ti and 68.5% Sn, while the constituents of the Ti$_6$Sn$_5$ phase are concerned. The dark, clearly limited zones are formed by almost pure Ti. The composition of SnAg$_{3.5}$Ti$_4$(Ce,Ga) solder in the selected spots is shown in **Table 3**.

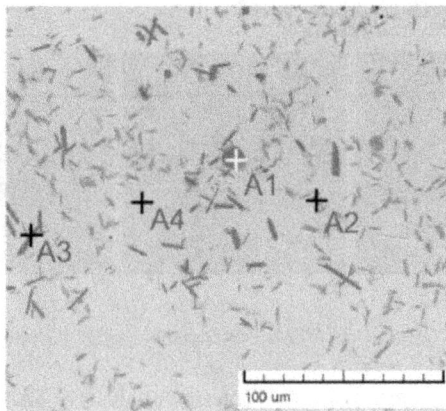

Figure 8.
Microstructure of the binary Ti-Sn system (SEM).

	Ti [wt. %]	Sn [wt. %]
A1	35.2	64.8
A2	4.36	95.64
A3	13.8	86.2
A4	0	100

Table 2.
Solder composition.

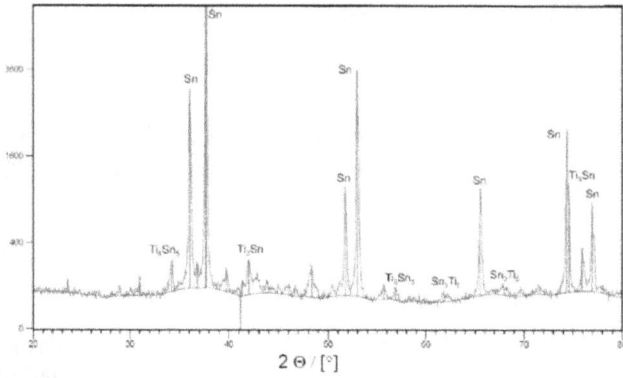

Figure 9.
X-ray record of SnTi$_2$ solder.

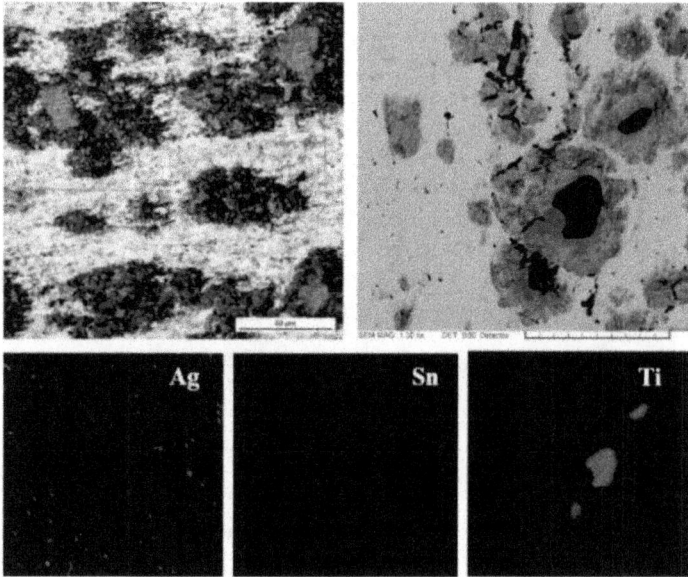

Figure 10.
Microstructure of SnAg$_{3.5}$Ti$_4$(Ce,Ga) solder (SEM) + concentration profiles of elements.

Figure 12 shows the record of the diffraction analysis of SnAg$_{3.5}$Ti$_4$(Ce,Ga) solder. The following phases were revealed: Ti$_6$Sn$_5$, Ag$_3$Sn, Ag$_3$Ti and Ti$_2$Sn. The solder consists of a tin matrix with non-uniformly distributed constituents of intermetallic Ti-Sn phases and the constituents of Ag$_3$Sn and Ag$_3$Ti (93% Sn, 6.5% Ag, 0.5% Ti) phase. The dark-grey phases shown in **Figure 11** in average contain 31.5% Ti and 68.5% Sn, while the Ti$_6$Sn$_5$ phase is concerned. The dark zones are composed of 100% Ti.

2.3 Bi-In25Sn18 solder

Figure 13 shows the microstructure of bismuth-based BiIn25Sn18 solder of eutectic composition. The solder was manufactured in the form of cast ingot. It exerts a fine multi-crystalline structure. All phases are uniformly distributed in the solder, without any traces of formation of conglomerates and/or clusters causing the heterogeneity in the chemical composition of solder elements. Composition

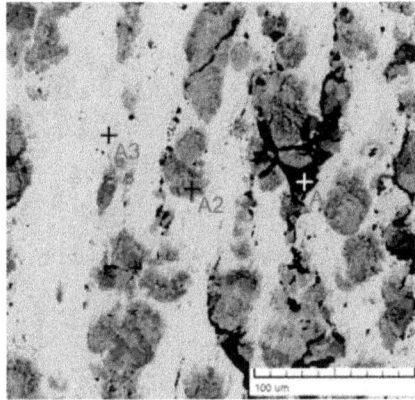

Figure 11.
The SnAg₃.₅Ti₄(Ce,Ga) solder (SEM).

	Ti [wt. %]	Ag [wt. %]	Sn [wt. %]
A1	99.15	0	0.85
A2	31.08	0.73	68.19
A3	0.5	6.56	93.14

Table 3.
$SnAg_{3.5}Ti_4(Ce,Ga)$ solder composition.

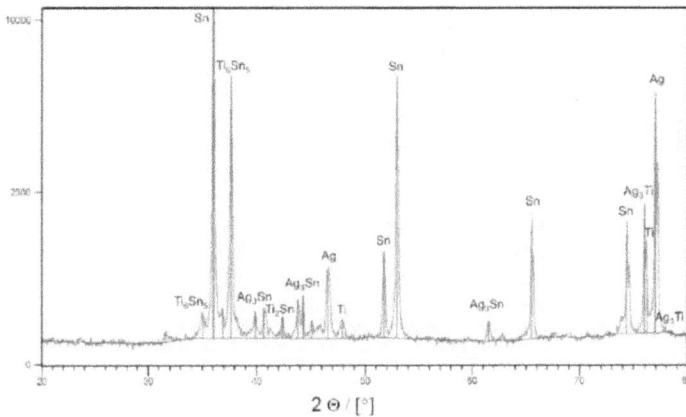

Figure 12.
X-ray record of $SnAg_{3.5}Ti_4(Ce,Ga)$ solder.

of the selected zones in BiIn25Sn18 solder is shown in **Table 4**. The ternary Bi-In-Sn system exerts two eutectics with a very low melting point. The composition of BiIn25Sn18 solder is very close to a ternary eutectics of Bi-In-Sn system, with a melting point of 77.5°C. Diffraction analysis performed on a specimen of BiIn25Sn18 solder (**Figure 14**) has revealed the presence of the following phases: Bi, Sn, In, $BiIn_2$, In_3Sn, $InSn_4$, BiIn and Bi_3In_5. The solder exerts very fine and uniformly distributed multiphase structure. The grey matrix contains in average 80% Bi, 12.5% In and 7.5% Sn. The white phase is composed of Bi-In (32.5% In and 67.5% Bi) constituents, while the Bi-In phase is concerned. The dark-grey constituents in **Figure 13** show an increased tin content (13.5% Bi, 6.5% In, 80% Sn).

35

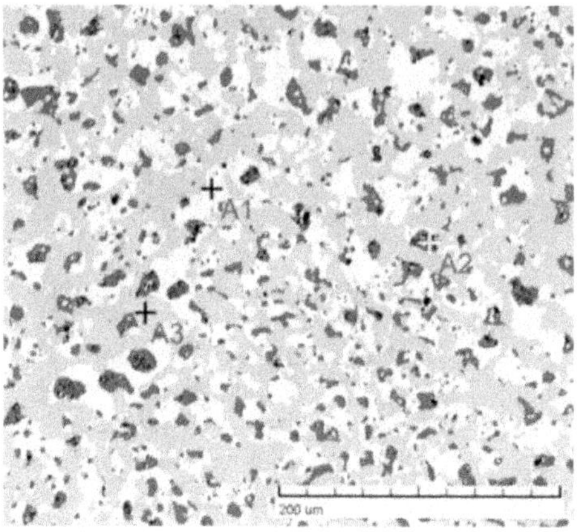

Figure 13.
The BiIn25Sn18 solder (SEM).

	Bi [wt. %]	In [wt. %]	Sn [wt. %]
A1	80.55	12.3	7.15
A2	11.7	0	88.3
A3	64.5	35.5	0

Table 4.
BiIn25Sn18 composition.

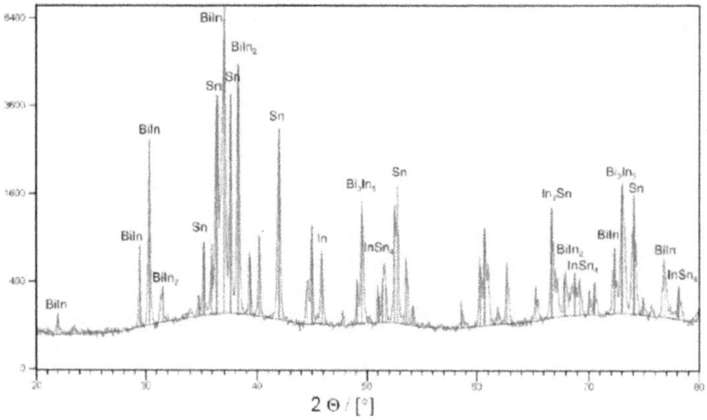

Figure 14.
X-ray record of BiIn25Sn18solder.

2.4 Results of shear strength

Figure 15 shows the values of shear strength of the analysed joints fabricated with $SnTi_2$, $BiIn_{25}Sn_{18}$ and $SnAg_{3.5}Ti_4(Ce,Ga)$ solders. The values give the average results of three measurements performed for each type of joint.

The highest strength on the copper and ceramic substrate was achieved with $SnAg_{3.5}Ti_4(Ce,Ga)$ solder. On the Al_2O_3 ceramics, it attained the strength of 33 MPa, while on the copper, it was 35 MPa. In the case of this solder, the difference between the shear strength values on the side of ceramics and the copper was smallest from amongst

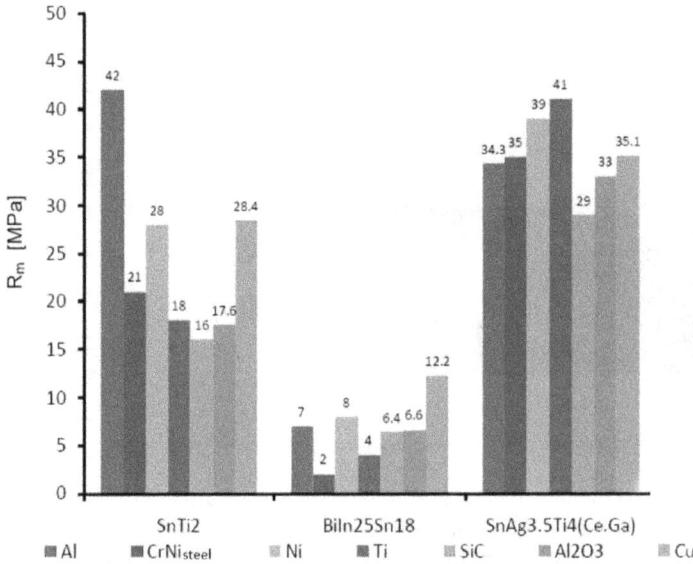

Figure 15.
Shear strength of SnTi$_2$, BiIn$_{25}$Sn$_{18}$ and SnAg$_{3.5}$Ti$_4$(Ce,Ga) solders.

all solders. The lowest shear strength values were attained with BiIn25Sn18 solder, both on the side of copper (12.2 MPa) and on the side of Al$_2$O$_3$ ceramics (6.6 MPa). The average shear strength values of SnTi$_2$ solder varied within the range from 16 to 42 MPa.

3. Analysis of soldered joints

3.1 Analysis of Al$_2$O$_3$-BiIn25Sn18 joint

Figure 16 shows a combined soldered joint of BiIn25Sn18-Al$_2$O$_3$ ceramics. A longitudinal crack was observed in the interface of BiIn25Sn18-Al$_2$O$_3$ joint, which has occurred along the entire specimen length. This crack was formed during the metallographic preparation of specimen due to poor plastic-elastic properties of the solder used.

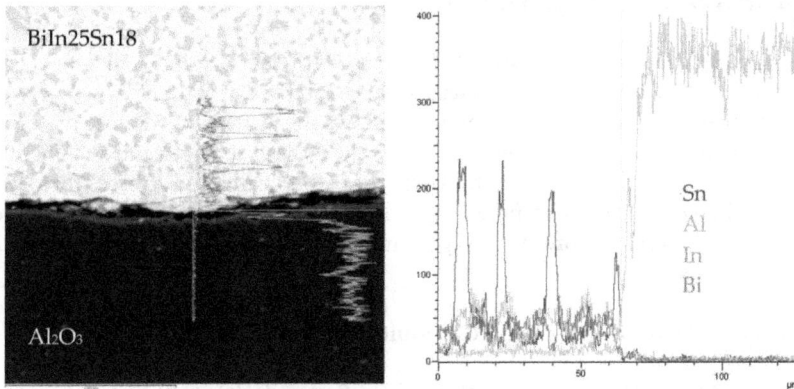

Figure 16.
Interface of Al$_2$O$_3$-BiIn25Sn18 joint (SEM) + concentration profiles of elements.

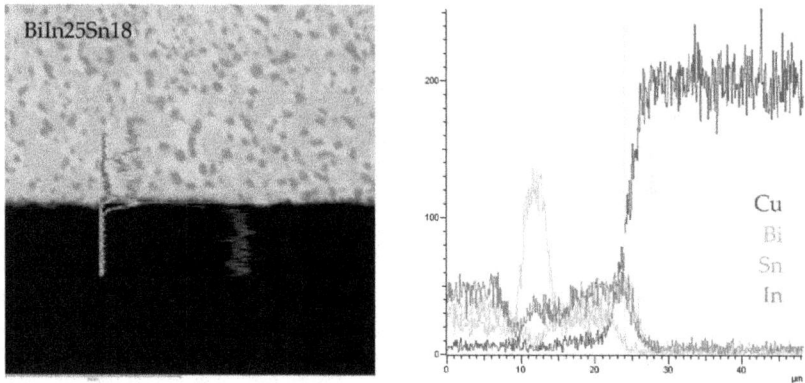

Figure 17.
Interface of Cu-BiIn25Sn18 joint (SEM) + concentration profiles of elements.

Based on the performed SEM and EDX analyses, it can be supposed that In primarily contributes to bond formation by creating indium oxide—In_2O_3. Besides In, Bi also partially contributes to bond formation. The presence of Sn is indifferent, and it does not exert any effect on the bond formation. Its presence was revealed in the grey phase. The interface contains exclusively a pale matrix composed of Bi and In. The formed reaction products are brittle with poor toughness that caused the crack formation in the bond plane.

3.2 Analysis of Cu-BiIn25Sn18 joint

Figure 17 shows the interface of BiIn25Sn18-Cu joint. In the interface of BiIn25Sn18 solder and copper, a noticeable increase in the proportion of the darker phase, formed mostly of Sn, may be seen. A continuous transition zone of reaction elements is formed in the solder interface. Based on the study of binary diagrams and performed analyses, formation of the following phases is supposed: Cu_3Sn, Cu_6Sn_5 and Cu_9In_4. A thinner, non-wettable phase rich in Cu (Cu_3Sn), which is followed with Cu_6Sn_5 phase, is formed in the joint interface. From the map of quantitative proportion of chemical elements, it is obvious that in the copper-solder interface mainly, the presence of In and Sn is exerted. However, Bi does not play any significant role in bond formation and does not create any phases with copper.

3.3 Analysis of Al_2O_3-$SnAg_{3.5}Ti_4(Ce,Ga)$ joint

Figure 18 shows the interface of Al_2O_3-$SnAg_{3.5}Ti_4(Ce,Ga)$ joint. A continuous reaction layer containing Ti was observed in the interface of Al_2O_3-$SnAg_{3.5}Ti_4(Ce,Ga)$ joint. This reaction layer allows the wetting of ceramic material and the tin-silver matrix of solder guarantees the desired strength and sufficient plastic properties of soldered joint to compensate the strains and stresses formed during cooling down. Slightly increased concentration of cerium was also observed in the joint interface. The presence of gallium and/or its effect upon bond formation was not observed (**Figure 19**).

3.4 Analysis of Cu-$SnAg_{3.5}Ti_4(Ce,Ga)$ joint

Figures 20 and **21** show the interface of Cu-$SnAg_{3.5}Ti_4(Ce,Ga)$ joint. A continuous layer of reaction elements is formed in the interface of solder and copper. Primary effect upon bond formation is exerted by Sn. Cu is dissolved in Sn matrix

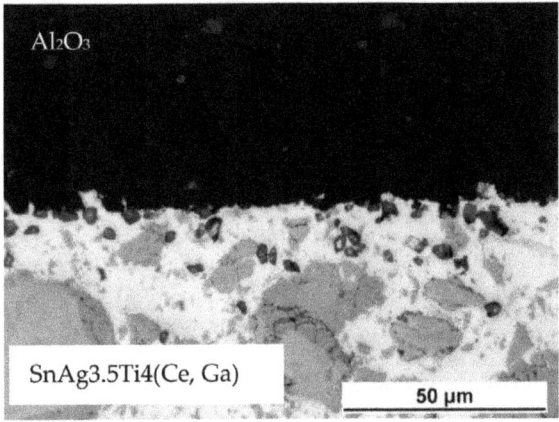

Figure 18.
Microstructure of interface of Al₂O₃-SnAg₃₅Ti₄(Ce,Ga) joint (SM).

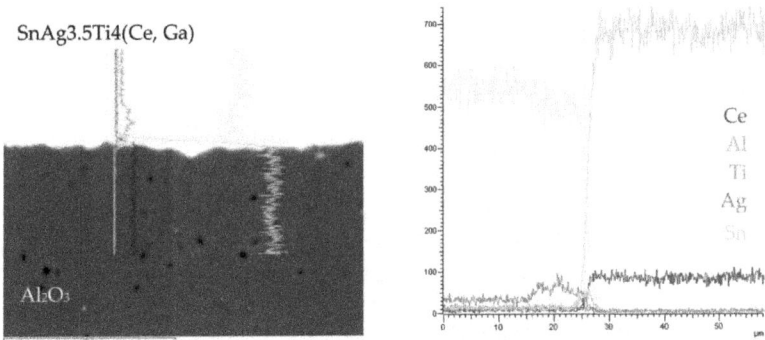

Figure 19.
Interface of Al₂O₃-SnAg₃₅Ti₄(Ce,Ga) joint (SEM) + concentration profiles of elements.

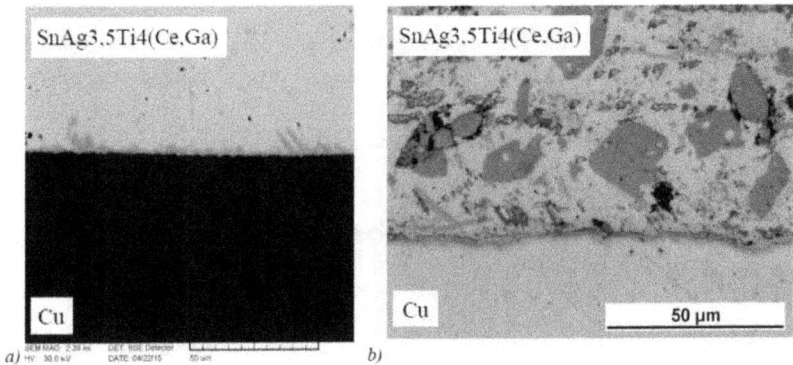

Figure 20.
Microstructure of interface of Cu-SnAg₃₅Ti₄(Ce,Ga) joint (a) (SEM) and (b) (SE).

and forms the Cu_3Sn and Cu_6Sn_5 phases which grow in the direction from the phase interface to solder matrix. Similarly to the case of soldering Al_2O_3, rapid dilution on the side of parent materials has not occurred. Sn matrix prevails in the solder, while the darker zones are formed by binary alloys of Sn with Ti and/or Ag.

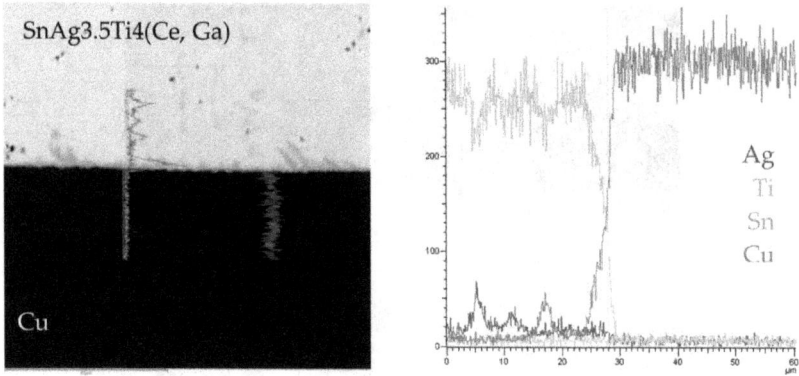

Figure 21.
Interface of Cu-SnAg$_{3.5}$Ti$_4$(Ce,Ga) joint (SEM) + concentration profiles of elements.

It was found out that Ti does not contribute at all in bond formation and its effect upon bond formation was unobservable. Ag element did not exert any significant interaction with the parent material; however, its presence in the interface was observed. The Ce and Ga elements occurred in the boundary in such low amounts that they could not be identified at all.

3.5 Analysis of Al$_2$O$_3$-SnTi$_2$ joint

Figure 22 shows the interface of SnTi2-Al$_2$O$_3$ joint. From the map of planar distribution of elements, it is obvious that the active Ti element significantly contributes in bond formation. It forms a continuous reaction layer, similarly as in the case of SnAg$_{3.5}$Ti$_4$(Ce,Ga) solder. This reaction layer is formed by Ti reaction with oxygen from ceramics at formation of titanium oxides, which alter the surface tension of ceramics and thus allow its wetting by solder. The effect of other elements (except Ti) upon bond formation was not observed (**Figure 23**).

3.6 Analysis of Cu-SnTi$_2$ joint

Figure 24 shows the interface of SnTi2-Cu joint. A continuous layer of reaction elements is formed in the interface of SnTi$_2$ solder and copper by dissolving Cu in

Figure 22.
Microstructure of interface of Al$_2$O$_3$-SnTi$_2$ joint (SE).

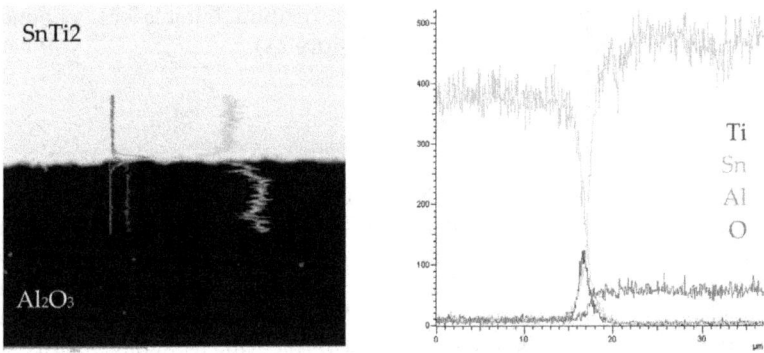

Figure 23.
Microstructure of interface of Al$_2$O$_3$-SnTi$_2$ joint (SEM) + concentration profiles of elements.

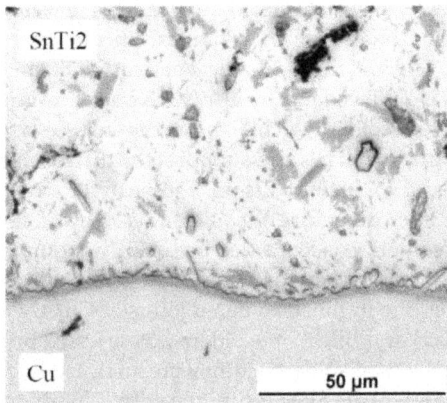

Figure 24.
Microstructure of interface of Cu-SnTi$_2$ joint (SE).

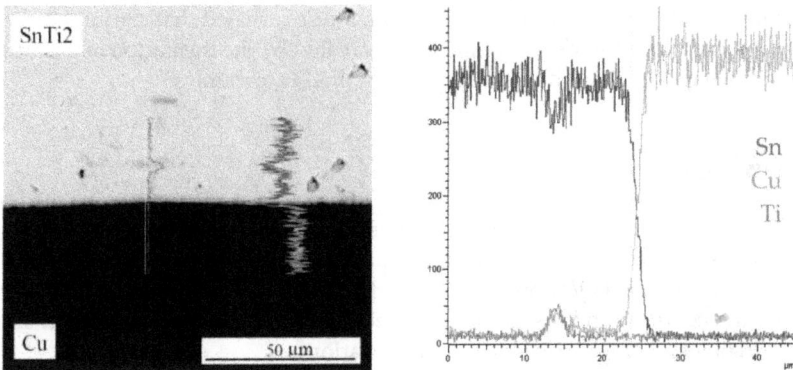

Figure 25.
Microstructure of interface of Cu-SnTi$_2$ joint (SEM) + concentration profiles of elements.

Sn matrix. We suppose the formation of similar Cu$_6$Sn$_5$ and Cu$_3$Sn phases as in the case of SnAg$_{3.5}$Ti$_4$(Ce,Ga) solder. The thickness of layer of intermetallic compounds depends on the level of soldering temperature and partially also on the dwell time at soldering temperature. Based on the records from EDW analyses, it may be concluded that Ti does not contribute in bond formation, but it is locally bound in the dark phases contained in solder matrix (**Figure 25**).

4. Conclusions

The aim of this chapter was to study the soldering of metallic and ceramic materials by the lead-free active solders based on Sn and Bi. Possibility of soldering ceramic materials is in considerable measure limited by the poor wettability of ceramic substrates with commercial solders at classical soldering technologies and owing to different thermal expansivity of soldered materials. Solderability study includes the application and subsequent study of soldering technology with ultrasound assistance applicable for ceramic materials and design of solder which allows to fabricate qualitatively acceptable soldered joint. From amongst the numerous methods used at present for joining ceramic materials, the technology of soldering with active solders was selected. This technology allows to wet both the metallic and also non-metallic materials as glass, ceramics, silicon, composites, etc. The power ultrasound was selected for mechanical activation of solders. Wetting of hard-to-wet materials is achieved just by ultrasound application, since it generates the cavitation in the liquid solder which disrupts the surface oxides, changes the surface energy of ceramic materials and supports the diffusion processes in the interface. The solders and soldered joints were subjected to a wide scope of analyses and experiments. The microstructure of solders was assessed in an initial state. The phase composition of solders was identified by the diffraction analysis.

Also static shear test was ranked to the tests of technological solderability of soldered joints. A series of combined soldered joints of Cu/Al_2O_3, fabricated with $SnTi_2$, $SnAg_{3.5}Ti_4(Ce,Ga)$ and $BiIn_{25}Sn_{18}$ solders was assessed by the performed experiments. The interfaces of soldered joint were analysed by the optical and scanning microscopy and by the SEM technique with EDX microanalysis. By the gradual selection, based on the desired properties, the soldering alloy type $SnAg_{3.5}Ti_4(Ce,Ga)$ was finally identified as the most perspective solder. This solder exerts a narrow melting interval from 221.4 to 224.6°C. The attained tensile strength was 53 MPa, whereas the shear strength varies within the range from 29 to 45 MPa. Regarding the mechanism of bond formation, it was revealed that the joint between the $SnAg_{3.5}Ti_4(Ce,Ga)$ solder and substrate is created by the formation of a continuous reaction layer of Ti with the surface layers of Al_2O_3 ceramics.

Acknowledgements

This work was supported by the Slovak Research and Development Agency under the contract no. APVV-17-0025. The paper was also prepared with the support of the VEGA 1/0089/17 project: Research of new alloys for direct soldering of metallic and ceramic; and Institutional Project SPAJKA: Investigation of new active lead-free solder alloy for space industry applications.

Soldering by the Active Lead-Free Tin and Bismuth-Based Solders
DOI: http://dx.doi.org/10.5772/intechopen.81169

Author details

Roman Koleňák*, Martin Provazník and Igor Kostolný
Faculty of Materials Science and Technology, Slovak University of Technology, Trnava, Slovak Republic

*Address all correspondence to: roman.kolenak@stuba.sk

IntechOpen

References

[1] Koleňák R, Žúbor P. Active Tin-Based Solder For Ceramic-Metal Joints. Materials Engineering. 2003;**3**(1) ISSN 1335-9053

[2] John J, Stephens K, Scott W. Brazing and Soldering. In: Proceedings of the 3rd International Brazing and Soldering Conference; ASM International; 2006. 428 p. ISBN-13: 978-0871708380

[3] Schwartz M. Handbook of structural ceramics. Mcgraw-Hill; 1991. 891 p. ISBN-10: 0070557195

[4] Chang SY. Active soldering of ZnS-SiO$_2$ sputtering targets to copper backing plates using an Sn$_{56}$Bi$_4$Ti(Ce,Ga) filler. Materials and Manufacturing Processes. 2006;**21**(8):761-765

[5] Hillen F, Pickart-Castillo D, Rass JI, Lugscheider E. Solder alloys and soldering processes for flux-free soldering of difficult- to-wet materials. Welding and Cutting. 2000;**52**(8):162-165

[6] Ruža V, Kroupová J. Characteristics of Ag$_{72}$CuTi active solder to soldering of ceramics with metals. Zváračské správy (Welding Reports) VÚZ. 1991;**41**(1):13-20

[7] Turwitt M. Joining Ceramics, Glass and Metal. Dusseldorf: DVS; 1997

[8] Inždinský K et al. Structure analysis of metal solder. Technology. 1995:54-56

[9] Feng KY et al. Brazing sapphire/sapphire and sapphire/copper sandwich joints using Sn-Ag-Ti active solder alloy. In: Solid State Phenomena. Trans Tech Publications; 2018. pp. 187-193

[10] Yu W, Liu Y, Liu X. Spreading of Sn-Ag-Ti and Sn-Ag-Ti(-Al) solder droplets on the surface of porous graphite through ultrasonic vibration. Materials & Design. 2018;**150**:9-16

[11] Yu W-Y et al. Wetting behavior in ultrasonic vibration-assisted brazing of aluminum to graphite using Sn-Ag-Ti active solder. Surface Review and Letters. 2015;**22**(03):1550035

[12] Qu W, Zhou S, Zhuang H. Effect of Ti content and Y additions on oxidation behavior of SnAgTi solder and its application on dissimilar metals soldering. Materials & Design. 2015;**88**:737-742

[13] Tsao L-C et al. Active soldering of aluminum–graphite composite to aluminum using Sn$_{3.5}$Ag$_4$Ti$_{0.5}$Cu active filler. International Journal of Materials Research. 2016;**107**(9):860-866

[14] Koleňák R. Soldering of ceramic materials. Zvárač (Welder). 2006;**3**(4)

[15] Koleňák R, Turňa M. Soldering of ceramic materials by active solders. Zváranie (Welding). 2001;**50**(3-4):75-78

[16] Chang SY, Chuang TH, Yang CL. Low temperature bonding of alumina/alumina and alumina/copper in air using an Sn$_{3.5}$Ag$_4$Ti(Ce Ga) filler. Journal of Electronic Materials. 2007;**36**(9):1193-1198

[17] Chang SY, Tsao LC, Chiang MJ, Tung CN, Pan GH, Chuang TH. Active soldering of indium tin oxide (ITO) with Cu in air using an SnAg$_{3.5}$Ti$_4$(Ce,Ga) filler. Journal of Materials Engineering and Performance. 2003;**12**(4):389

[18] Kuper W et al. Phase formation and reaction kinetics in system. Ti Sn Metallkde. 1998;**89**:855-862

[19] Xian AP. Joining of sialon ceramics by Sn-5 at % Ti based ternary active solders. Journal of Materials Science. 1997;**32**:6387-6393

[20] S-Bond. Euromat, Industrial Surface Solutions [Internet]. Available from: https://www.euromat.de/ [Accessed: 25.08. 2010]

[21] Rigó P. Soldering of metallic and non-metallic materials using mechanical [thesis]. STU MTF Trnava; 2003

[22] Koleňák R. Physical-metallurgical aspects of soldering ceramics with metals [dissertation]. STU MTF Trnava; 2001

[23] Brochu M et al. Joining silicon nitride ceramic using a composite power as active. Materials Science and Engineering. 2004;**374**

[24] Švehla Š, Figura Z. Ultrasound in Technology. 2007:151. ISBN 978-80-227-2705-1

[25] Hanuz Ltd. Some Words about Ultrasound [Internet]. Available from: http://www.hanuz.sk

[26] Suslick K. The chemical Effects of Ultrasound [Internet]. Available from: http://www.scs.illinois.edu/suslick/documents/sciamer8980.pdf [Accessed: 11.04. 2010]

Chapter 4

Room-Temperature Formation of Intermixing Layer for Adhesion Improvement of Cu/Glass Stacks

Mitsuhiro Watanabe and Eiichi Kondoh

Abstract

Reliable and high-precision Cu/glass stacks are particularly desirable for micro-electromechanical systems and packaging technologies. One solution for improving the adhesion strength of Cu/glass stacks is to form adhesion layers between the Cu films and the glass substrate. Many studies have shown that a strong adhesion layer is formed at the interface by high-temperature annealing when a Cu alloy is used instead of pure Cu. It is important to reduce the temperature and process time in order to reduce the thermal budget and fabrication cost. Therefore, the room-temperature process for fabrication of Cu/glass stack is desirable. In this chapter, typical advanced low-temperature processes including room-temperature process are introduced.

Keywords: adhesion improvement, low-temperature process, room-temperature process, intermixing, Cu/glass stack

1. Introduction

In microelectromechanical system (MEMS) and packaging technologies, high reliability of Cu metallization of glass substrates is strongly required. In the field of structural materials, fusion welding such as arc welding and laser welding is usually applied for dissimilar-metal or dissimilar-material welding, but in these welding methods, extremely high energy are needed for melting of metals, and it is difficult for joining of micrometer-scale or nanometer-scale precision. Recently, solid-state welding methods such as friction stir welding and magnetic pulse welding are developed. These methods have no high energy comparing with the fusion welding because these methods are achieved for joining at solid state. However, atomic diffusion for achievement of joining needs to be accelerated at solid state. For acceleration of the atomic diffusion, friction is usually generated between the welding materials. Therefore, brittle materials such as glass and silicon wafer are broken when the solid-state welding methods are applied for the MEMS stacks. Also, these fusion welding and solid-state welding methods usually produce thick brittle intermixing layer such as intermetallic compound with a scale of micrometers or larger. The brittle layer produced at the interface lowers the mechanical strength of the joint. Therefore, atomically scaled interface bonding is demanded for fabricating a highly reliable Cu/glass structure.

One of the solutions for strengthening the interface bonding is the formation of an adhesion layer between a Cu film and a glass substrate. It is a common sense in vacuum engineering to insert a reactive metal such as Al or Ti between Cu and glass. However, this technique is not very useful in three-dimensional MEMS/packaging, because these metals must be deposited with sequential high-vacuum deposition methods, which do not give a good step coverage. In addition, reactions between Cu and those metals can lead to a significant increase in resistance when the Cu film is thin.

Koike et al. have investigated the interfacial properties of the annealed Cu-Mn/ glass structure and reported that adhesion improvement was observed by formation of a several nm thick Mn oxide layer at the interface [1, 2]. Yi et al. reported the formation of interfacial layer by annealing in Cu-Mg/glass and showed that the adhesion strength improved by formation of a Mg oxide layer at the interface [3]. Other elements, such as Al, Ti, and Cr, added to Cu were studied previously, and they were reported to improve the adhesion strength between Cu and various substrates (not only glass) [4–6]. These studies mentioned above indicate that the effective adhesion layers contain elements that are easily oxidizable and miscible in Cu. However, it should be noted that these studies required heat treatment during/after deposition. For achieving the general trend of temperature reduction during microelectronic fabrication, room-temperature or lower-temperature adhesion improvement is required.

2. Low-temperature bonding using ion beam etching

For bonding at lower temperature, surface refresh is one of the effective methods, because contamination such as oxide scale and inclusion is formed at material surface exposed in the atmosphere. The contamination usually prevents from bonding of the materials. High-energy ion beam irradiation is useful for cleaning of the contaminated material surface. When the ion beam irradiation and the bonding of materials having the refreshed surface are done at same high-vacuum environment (without exposing atmosphere), bonding of both activated surfaces is achieved without suing a high temperature. This method is usually called "surface activation method" (**Figure 1**) [7].

Research group of Suga has studied several combination of similar- and dissimilar-material bonding (Si/Si [8, 9], Al/Al [10], Cu/Cu [8], Si/SiO$_2$ [11]) using the surface activation method. Especially, aluminum always has strong oxide scale at the surface, but smooth and clear bonding interface without voids is formed by using this method (**Figure 2**) [10]. This indicates that ion beam irradiation is an effective

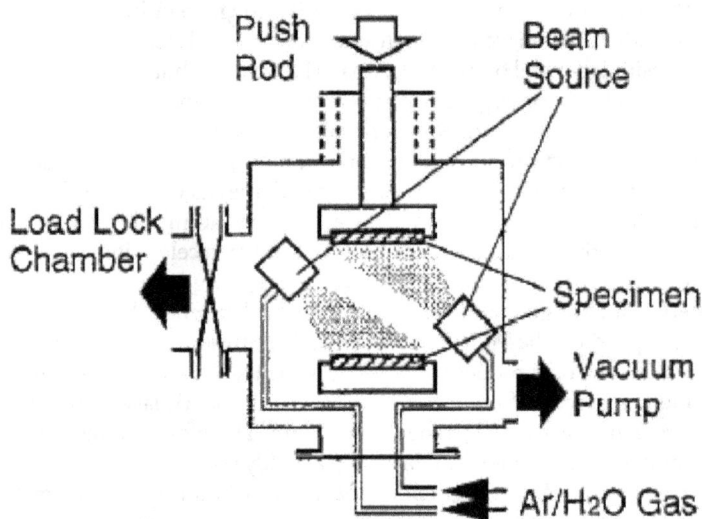

Figure 1.
Surface activation bonding method [11].

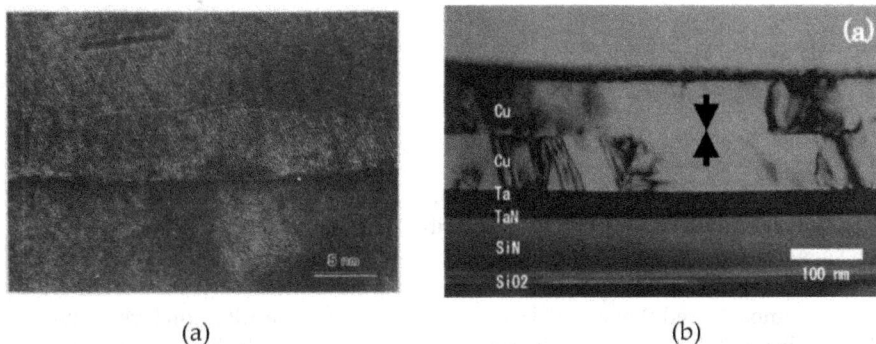

Figure 2.
TEM images of (a) Al/Al [10] and (b) Cu/Cu [8] interfaces fabricated by surface activation bonding.

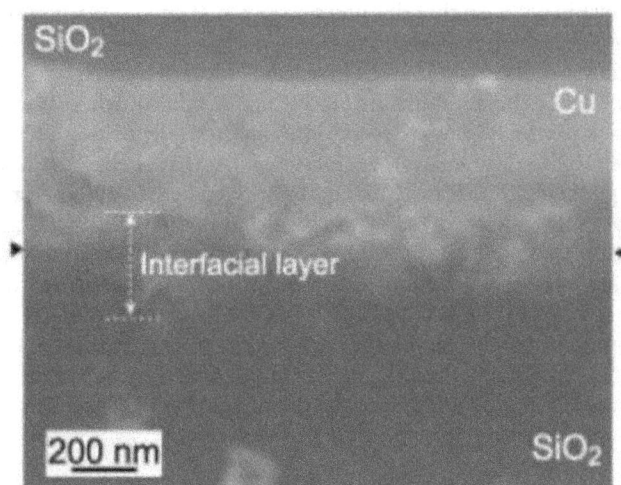

Figure 3.
Cu/SiO₂ interface fabricated by surface activation bonding [12].

method for removing the oxide scale. Also, achievement of bonding of Si and Si suggests that this method can be used for brittle materials. In addition, Takagi et al. reported the bonding of Si and SiO_2 using the surface activation method [11]. This indicates that this method can be achieved by the dissimilar-material bonding. A report on Cu/glass (Cu/SiO_2) bonding using the surface activation method also existed [12]. When the bonding temperature is increased to 423 K, good adhesion is obtained (**Figure 3**). The process temperature is very low than that of conventional diffusion bonding. This is considered to indicate that the surface cleaning by the ion beam irradiation is effective for the lower-temperature bonding.

3. Room-temperature formation of adhesion layer utilizing for adhesion improvement of Cu/glass stacks

As mentioned in Section 1, formation of adhesion layer at Cu/glass interface is an effective method for improvement of adhesion. Recently, we demonstrate room-temperature formation of Cu/glass stack with high adhesion strength [13, 14]. This adhesion improvement is due to effect of ZnO-based adhesion layer formed at room temperature. The formation process is simple and affordable and is similar to that of electroless plating (ELP) of Cu. The research process leading to the development of this fabrication process is described below.

ZnO has gained considerable attention in microelectronics as an alternative transparent conductor [15, 16], because Zn is a recyclable, abundant, and affordable element. ZnO can be deposited by various techniques such as sputtering (SPT) [17, 18], sol-gel [19], chemical vapor deposition [20, 21], and supercritical fluid chemical deposition [22, 23]. In addition, Zn is miscible in Cu up to 38.27 at% Zn [24], and ZnO generally shows good adhesion to glasses or oxides [25, 26], indicating that ZnO is a potential adhesion layer in Cu/glass structures. Indeed, past research studies demonstrated that a ZnO layer works as an effective adhesion layer between Cu and glass [27, 28]. However, the films employed in those studies had micrometer thicknesses and a high-roughness surface topography, obviously inappropriate for micro-/nanoelectronic applications. Recently, we demonstrated that thin ZnO layers improve the adhesion between Cu and glass.

The relationship between the adhesion strength evaluated by a microscratch tester following JIS R3255 specifications and the Cu deposition method is shown in **Figure 4**. No delamination was observed in stack in which Cu was deposited by electroless plating (ELP) on a ZnO/glass substrate (ELP-Cu/ZnO/glass), whereas the adhesion strength of stack in which Cu was deposited using vapor deposition (VD) on a ZnO/glass substrate (VD-Cu/ZnO/glass) was low as well as that of the Cu/glass structure. Note that the glass substrate is fractured at the applied load of 500 mN; namely, the fracture of glass substrate occurred before the delamination in ELP-Cu/ZnO/glass stack. This means that the ELP-Cu/ZnO/glass stack has excellent adhesion.

Figure 5 shows the cross-sectional scanning transmission electron microscope (STEM) images and STEM-EDX (energy dispersive X-ray spectrometer equipped with STEM) maps of (a) the VD-Cu/ZnO/glass and (b) ELP-Cu/ZnO/glass stacks. In the VD-Cu/ZnO/glass stack, the Cu film, ZnO layer, and glass substrate were separately observed. At the Cu/ZnO interface, several voids were formed, as indicated by

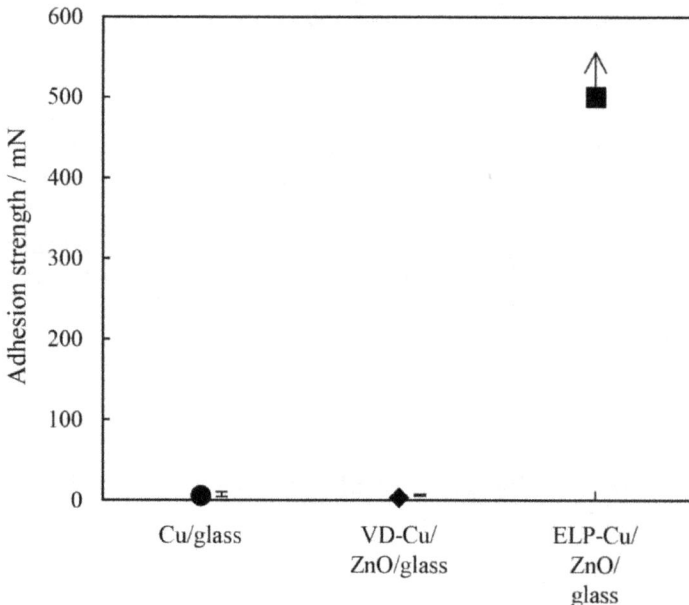

Figure 4.
Relationship between adhesion strength and deposition method.

Figure 5.
Cross-sectional STEM images and STEM-EDX maps of (a) VD-Cu/ZnO/glass and (b) ELP-Cu/ZnO/glass stacks.

arrows in STEM image, indicating that the low adhesion strength is due to no intermixing at each interface. In contrast, the ELP-Cu/ZnO/glass stack exhibited a smooth interface, but no clear ZnO layer was observed. Careful observation revealed that the formation of an approximately 10-nm-thick layer was produced at Cu/glass interface. Strong Cu signals were obtained over the entire ELP-Cu film region, whereas a slight low Cu intensity was observed in the 10-nm-thick layer. Zn was detected at all regions of Cu film, 10-nm-thick layer, and glass substrate. Pd is a catalyst element used in the ELP, and it was observed in the 10-nm-thick layer. This indicates that Pd diffused into the ZnO layer during the ELP process, losing its original particulate shape (**Figure 6**). O and Si signals were detected in the 10-nm-thick layer. That is, the 10-nm-thick layer consists of Cu, Pd, Zn, O, and Si. The formation of such an intermixing layer at the Cu/glass interface significantly improved the adhesion.

When the VD process (processed at room temperature) was used to deposit the Cu films, an intermixing layer was not formed, and the ZnO layer was clearly observed, and voids were formed at the Cu/ZnO interface. On the other hand, it should be considered that the intermixing layer was formed during the ELP process, which was carried out at a deposition/plating temperature of almost room temperature (308 K). Also, ZnO layer was removed after Cu electroless plating. In addition, the SPT-Cu/ZnO/glass stack has also high adhesion strength like the ELP-Cu/ZnO/glass stack when pretreatment of Cu electroless plating was applied to the ZnO/glass substrate. In general, high temperatures were used to form such a reaction layer. The accelerated diffusion reaction is considered to be due to effect of catalytic role of Pd. We assume that the thinning of the ZnO layer decreased the diffusion distance of Cu and Si, which enhanced the Pd-promoted intermixing of Cu, Zn, O, and Si.

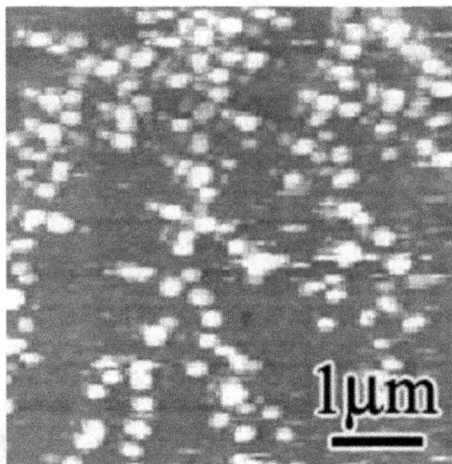

Figure 6.
Surface morphology of Pd-catalyzed ZnO/glass stack.

As mentioned above, the formation of intermixing layer at Cu/glass interface was found to result in adhesion improvement of Cu/glass stack. The intermixing layer is formed at room temperature, but reaction between Cu and glass does not occur at room temperature in general. For understanding the formation mechanism of intermixing layer at room temperature, the interfacial reaction at each process is necessary to be investigated.

At first, in order to confirm the incorporation of Zn into the glass (SiO_2) surface, the interfaces of ZnO/SiO_2 were examined before removing the ZnO layer [14]. This is because ZnO layer deposited at 673 K by metal organic chemical vapor deposition (MOCVD) was used in the results shown so far. **Figure 7** shows transmission electron microscope (TEM) images and Zn, Si, and O maps of the interfaces of (a) as-deposited MOCVD-, (b) as-deposited MBE-, and (c) annealed MBE-ZnO/ glass stacks. Deposition temperatures of MOCVD- and MBE-ZnO were 673 K and room temperature, respectively. The MBE-ZnO/glass stack was then annealed at 623 K for 60 min. In the as-deposited MOCVD- and the annealed MBE-ZnO/SiO$_2$ stacks, approximately 5-nm-thick layer formation was observed at SiO_2 side of the ZnO/SiO_2 interfaces, whereas no reaction layer was observed at the as-deposited MBE-ZnO/SiO$_2$ interface. The Zn and Si signals were observed in the interfacial layer regions in the as-deposited MOCVD- and the annealed MBE-ZnO/SiO$_2$ stacks, whereas the as-deposited MBE-ZnO/SiO$_2$ stack showed very sharp transition between the Zn and Si intensities. These results indicate that Zn was diffused approximately 5 nm into the SiO_2 when the stacks were processed at high temperatures; i.e., the SiO_2 surface was doped with Zn by heat treatment during or after ZnO deposition.

Figure 8 shows a high-resolution TEM (HRTEM) image of the ZnO/SiO_2 interface. A typical amorphous structure having no fringe contrast can be observed in the glass substrate region. The lattice image is clearly observed in the interfacial layer regions as in the ZnO layer region, obviously proving that the interfacial layer is crystalline. **Figure 9(a)** shows a high-angle annular dark-field (HAADF) image of the ZnO/SiO_2 interface. The interfacial layer was composed of high-contrast regions (A in **Figure 9(b)**) and low-contrast regions (B in **Figure 9(b)**). TEM-EDX revealed that the A-region is Zn-containing SiO_2 ($SiO_2(Zn)$) and the B-region is an equilibrium Zn_2SiO_4 phase (**Table 1**).

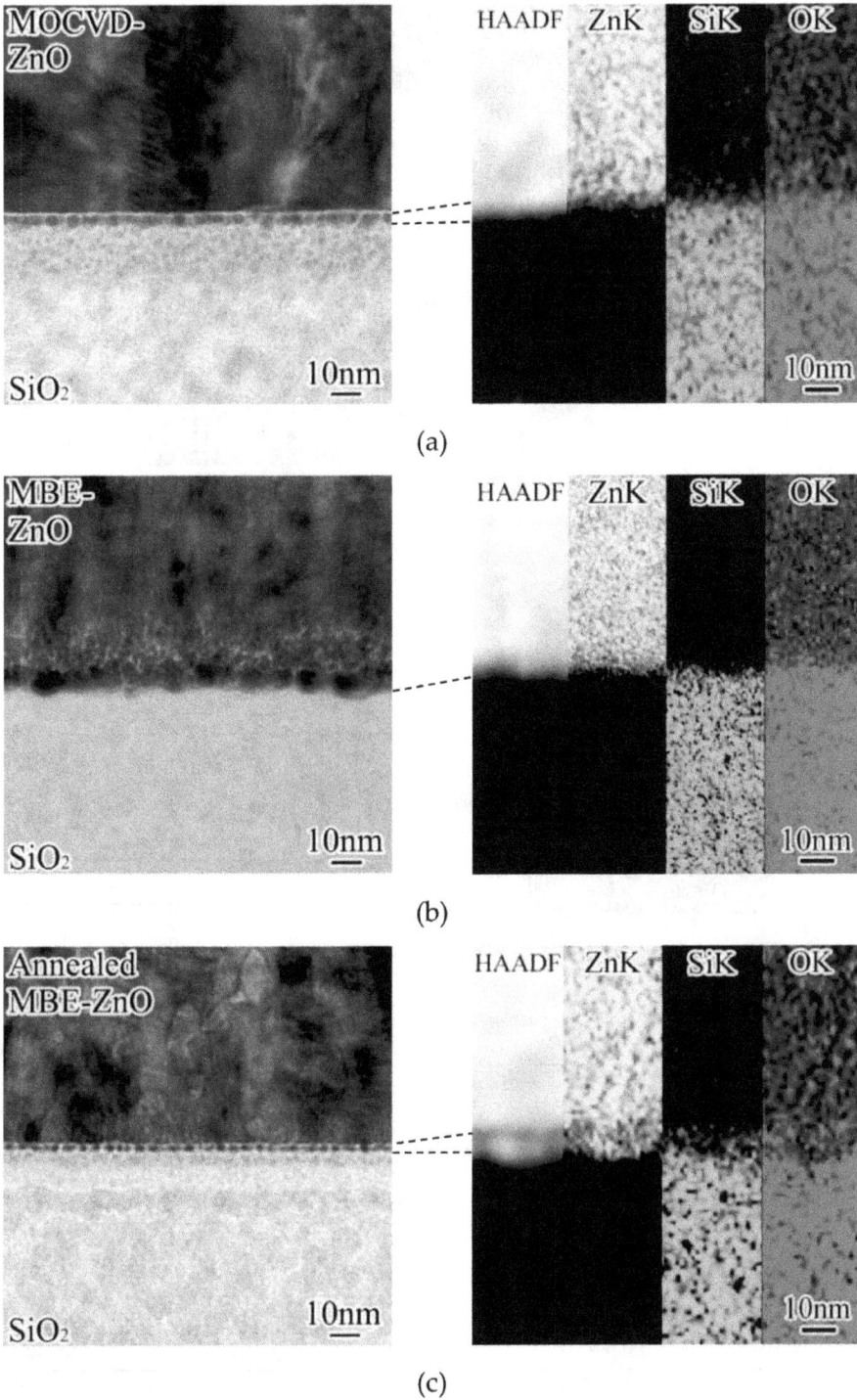

Figure 7.
TEM images and TEM-EDX maps of the interfaces of (a) as-deposited MOCVD-, (b) as-deposited MBE-, and (c) annealed MBE-ZnO/SiO₂ stacks. MBE means molecular beam epitaxy.

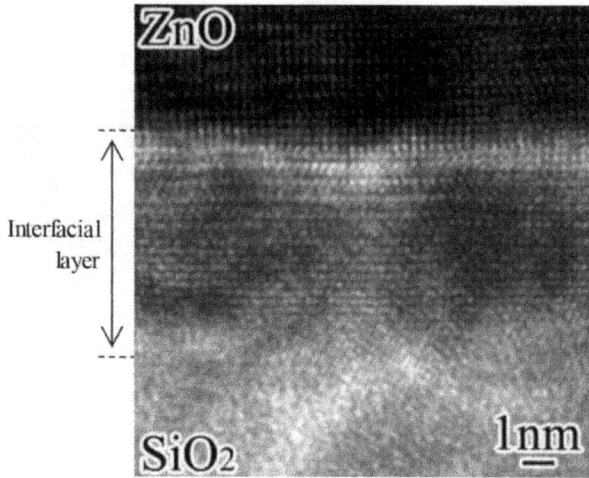

Figure 8.
HRTEM image of ZnO/SiO₂ interface.

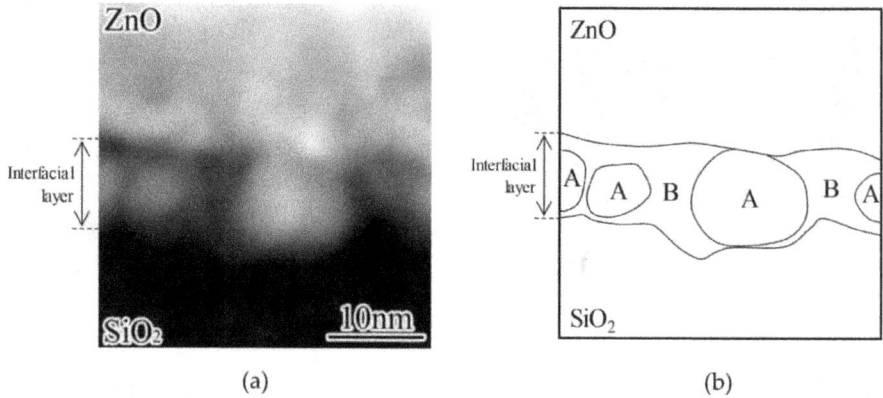

(a) (b)

Figure 9.
(a) HAADF image of the ZnO/SiO₂ interface and (b) schematic illustration of (a).

	Atomic percentage		
	Zn	Si	O
A-region (higher-contrast region)	42	18	40
	46	16	38
	36	21	43
B-region (lower-contrast region)	27	17	56
	29	19	52
	33	17	50

Table 1.
TEM-EDX results of higher- and lower-contrast regions formed in interfacial layer.

The effect of dip to acid solution of ZnO/SiO₂ substrate was investigated. The dip to acid solution of substrate is surface treatment for substrate in conventional pretreatment of electroless plating. This process results in removing the ZnO layer.

TEM observation of the ZnO/glass stack after removing ZnO layer revealed the presence of an extremely thin surface layer (**Figure 10(a)**). Zn, Si, and O maps indicate that the surface layer consists of Zn, Si, and O (**Figure 10(b)**), and this composition distribution was similar to that of the interfacial layer formed at the ZnO/SiO$_2$ interface by heat treatment. This means that the interfacial layer remained at the SiO$_2$ surface as a Zn-doped layer even after the original thick ZnO layer being removed.

The relationship between noble metal particle and Zn-surface-doped layer was investigated. The noble metal particles (Pt in this case) and Cu films were deposited on the Zn-surface-doped SiO$_2$ and non-doped SiO$_2$, and the adhesion strength was investigated. When the Zn-surface-doped SiO$_2$ was used, remarkable adhesion improvement (>500 mN) was observed, whereas adhesion improvement was insufficient (26 mN) when the nondoped SiO$_2$ was used. This indicates that the doping of SiO$_2$ surface with Zn was effective for improving the adhesion at Cu/glass interface.

In order to know the effect of the noble metal particle on adhesion improvement, Cu was sputtered on the Zn-doped SiO$_2$ with/without Pt, Pd, or Pt-Pd particles, and the adhesion strength was evaluated. Remarkable adhesion improvement (>500 mN) was observed when Pt, Pd, or Pt-Pd particles were employed. This proves that Pt, Pd, and Pt-Pd particles are effective catalysts and, moreover, the combination of the Zn dope and the noble metal catalyzation enhances the adhesion between Cu films and SiO$_2$ substrate.

Figure 11 shows cross-sectional TEM image and EDX maps of the Cu/Pt/Zn-doped SiO$_2$ stack. The results were similar to that of the Cu/Pd/Zn-doped glass stack (**Figure 5(b)**). An extremely thin intermixing layer was clearly formed at the interface, and its composition was Cu$_{42}$Pt$_{18}$Zn$_{0.8}$Si$_{15}$O$_{22}$. From these observations, it is safely said that the formation of this intermixing layer results in the adhesion improvement. In the previous study, the formation of such an intermixing layer was observed when we used Pd as a catalyst, which shows the intermixing acceleration [13].

The effect of the kind of glass substrate on adhesion strength was investigated [29]. Each stack was fabricated by using each glass substrate (borosilicate glass, soda glass, SiO$_2$, SiO$_2$ thermal oxide growth on Si), SPT-ZnO layer, Pt as a catalyst, and SPT-Cu deposited at room temperature. Adhesion strengths were higher than 500 mN in any stacks, and no film delamination occurred even when any glass substrates were used. This means that Zn-doped layer was formed at each glass surface and that intermixing at room temperature occurred at Cu/glass interface. It is found that this adhesion enhancement is not influenced by the kind of glass substrate.

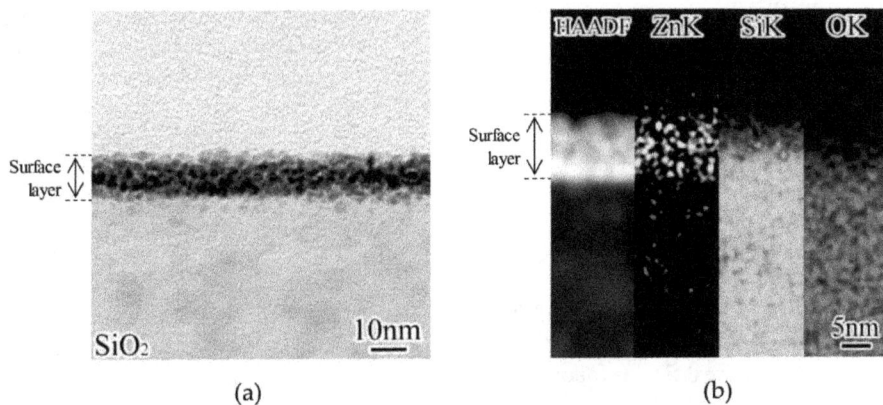

Figure 10.
(a) TEM image and (b) TEM-EDX maps of SiO$_2$ surface after the removal of ZnO layer.

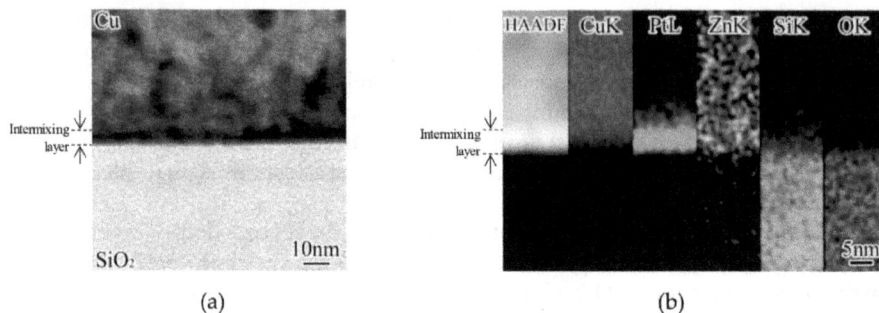

Figure 11.
(a) TEM image and (b) TEM-EDX maps of the interface of a Cu/Pt/Zn-doped SiO₂ stack.

The influence of the kind of ZnO layer on adhesion strength was examined [30, 31]. ZnO layer was fabricated by sol-gel method, metal organic decomposition (MOD), and supercritical fluid chemical deposition (SFCD), and we investigated whether these ZnO layers improve the adhesion of Cu/glass stacks. Adhesion strengths of Cu/glass stacks fabricated by using sol-gel-ZnO or MOD-ZnO were higher than 500 mN. When SFCD-ZnO was employed, adhesion strength (21 mN) was almost 20 times stronger than that of the Cu/non-treated glass stacks. The sol-gel- and MOD-ZnO layers were transparent as well as MOCVD- and MBE-ZnO layers, but the SFCD-ZnO layer had brown or white color. The color of the ZnO layer is considered to show crystallinity of ZnO. These results suggested that sol-gel-, MOD-, and SFCD-ZnO layers are effective for improving the adhesion of Cu/glass stacks and that the degree of improvement depends on the film quality.

From the above results, fabrication process of Cu/glass stack for significant adhesion improvement is shown in **Figure 12**. This process is simple and allows conventional Cu electroless plating or any other common Cu deposition methods, such as SPT and VD, to be used. **Figure 13** shows surface images of (a) VD-Cu/non-treated glass stack and (b) VD-Cu/glass stack fabricated using the process indicated in **Figure 12** after crosscut test (JIS K5600-5-6). It is clearly found that no Cu film delamination was observed in VD-Cu/glass stack fabricated using the process indicated in **Figure 12**, whereas Cu films were completely delaminated from glass substrate in VD-Cu/non-treated glass stack. The key in this process is the combination of Zn doping and noble metal catalyzation, which accelerates atomic intermixing at the Cu/glass interface. For obtaining the Zn-doped glass surface, annealing of the ZnO/glass stack during or after ZnO deposition was effective treatment.

Based on the observed results, interfacial reaction during fabrication process indicated in **Figure 12** is discussed. **Figure 14** shows schematic illustrations of the reaction at the ZnO/SiO₂ interface at 573–673 K. SiO₂(Zn) and Zn₂SiO₄ phases were produced at the interface (**Table 1**) by annealing of ZnO/SiO₂ stack at 623–673 K. This layer remained as the surface layer after ZnO removal. According to the ZnO/SiO₂ phase diagram [32], Zn₂SiO₄ is formed over 1573 K. This temperature is much higher than our process temperature. This indicates that the formed Zn₂SiO₄ phases were produced by diffusion reaction.

The reaction between ZnO and SiO₂ does not produce other compounds $(2ZnO + SiO_2 \rightarrow Zn_2SiO_4)$. Therefore, based on our observations, we formulated a possible overall reaction as follows:

$$4ZnO + 2SiO_2 \rightarrow Zn_2SiO_4 + SiO_2(2Zn) + 2O \qquad (1)$$

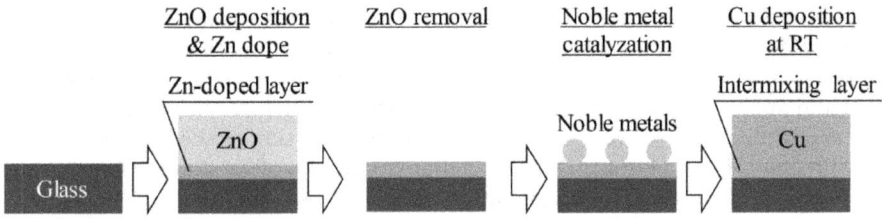

Figure 12.
Schematic diagrams of process for high-adhesion Cu/glass stack. RT means room temperature.

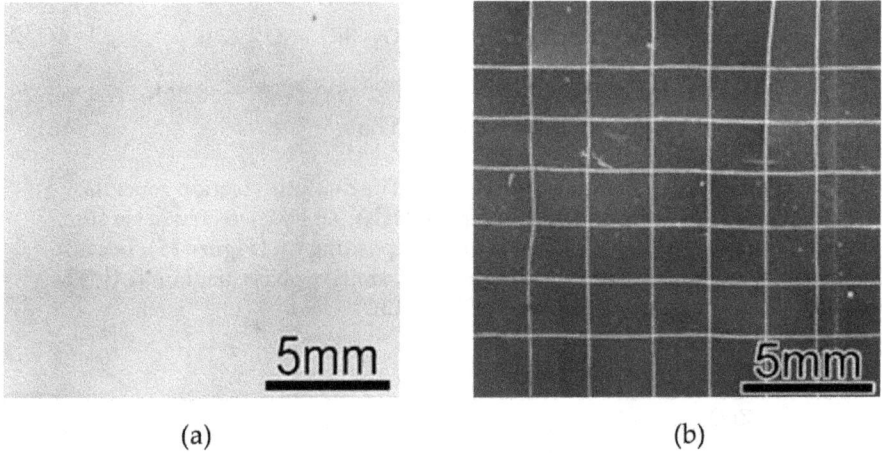

(a)   (b)

Figure 13.
*Macroscopic appearances of (a) VD-Cu/non-treated glass stack and (b) VD-Cu/glass stack fabricated by using the process for high-adhesion Cu/glass stack (**Figure 12**).*

where the coefficient of Zn is introduced as 2Zn to satisfy the stoichiometry of Zn. The bracket of Zn shows that Zn is free metal and dissolves in or coexists with the SiO_2.

The formula (1) indicates that the molar ratio of the formed Zn_2SiO_4 and the $SiO_2(Zn)$ is 1:1 and that more ZnO is consumed in the reaction than SiO_2. Indeed, from the TEM-EDX analyses, the volumes of these phases are almost identical apart from the molar density of these phases. The stoichiometry also shows that free Zn forms and dissolves in SiO_2. As a result, as observed (**Table 1**), the Zn_2SiO_4 and the $SiO_2(Zn)$ coexist at the interface.

Elementary reactions of reaction (1) can be expressed as follows:

$$ZnO + SiO_2 \rightarrow Zn_2SiO_4 \tag{2}$$

$$ZnO \rightarrow 2Zn + 2O \tag{3}$$

$$SiO_2 + 2Zn \rightarrow SiO_2(2Zn) \tag{4}$$

The above formulae simply indicate that ZnO is decomposed to Zn and O. Presumably, the formula (2) ignites these reactions to proceed. Once Zn diffuses into SiO_2, oxygen in ZnO becomes less and promotes oxygen diffusion.

Next, interfacial reaction of the formation of the intermixing layer is discussed. **Figure 15** shows schematic illustrations of the reaction at the Cu/Zn-surface-doped SiO_2 interface at room temperature. From the observation and analysis results, an extremely thin intermixing layer was formed at the interface at room temperature, and the composition of the layer was $Cu_{42}Pt_{18}Zn_{0.8}Si_{15}O_{22}$ (**Figure 11**). The composition ratio of Zn, Si, and O is Zn:Si:O = 2.1:39.7:58.2. This composition has a Si/O ratio close to 2:3. Therefore, we can safely say that this phase is an oxygen-deficient silicon oxide, presumably Si_2O_3 ($SiO_2 \cdot SiO$) that contains impurity Zn. The reactions can be expressed as the following formulae:

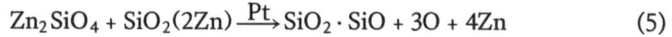

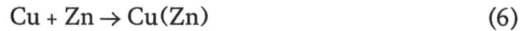

$$Zn_2SiO_4 + SiO_2(2Zn) \xrightarrow{Pt} SiO_2 \cdot SiO + 3O + 4Zn \tag{5}$$

$$Cu + Zn \rightarrow Cu(Zn) \tag{6}$$

The formula (5) is a Pt-catalyzed reaction. The catalytic reaction generates free oxygen and free Zn. As shown in **Figure 11(b)**, Zn was detected in Cu film. Therefore, the repelled Zn dissolves into the depositing Cu (**Figure 15**), because Zn is well miscible in Cu according to the Cu/Zn binary phase diagram [24]. The formula (6) expresses the dissolution of Zn in Cu.

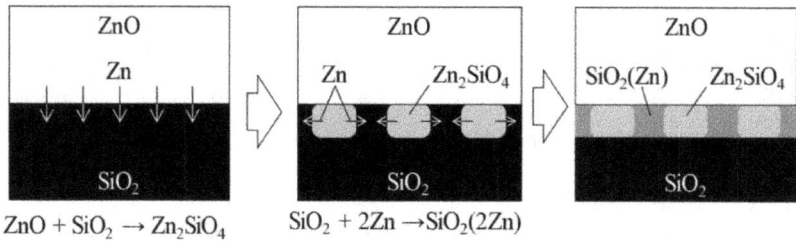

Figure 14.
Schematic illustrations of reaction at ZnO/SiO₂ interface at 573–673 K.

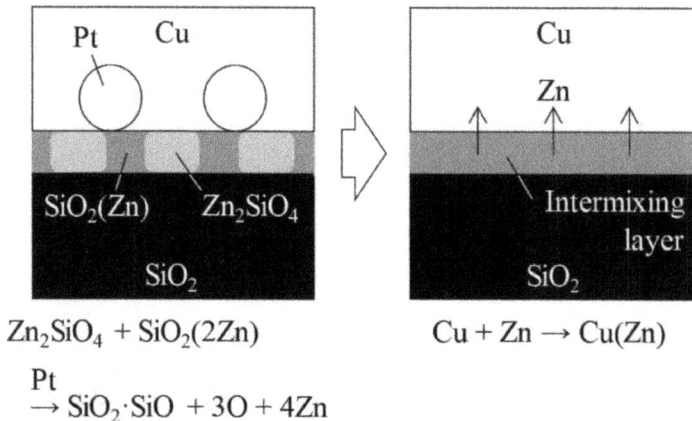

Figure 15.
Schematic illustrations of reaction at Cu/Zn-doped SiO₂ interface at room temperature.

4. Summary of this chapter and prospect for fabrication of the next-generation microelectronic devices

In this chapter, lower-temperature process for fabrication of high-adhesion Cu/glass stack was introduced. The surface activation bonding is one of the lower-temperature processes for similar- and dissimilar-material bonding. Also, room-temperature formation process of intermixing layer is also introduced. A nanoscale ZnO layer improves the adhesion strength between Cu and glass even if metallization was carried out at room temperature. A Cu/glass stack with a high adhesion strength is successfully fabricated by the combination of Zn dope and noble metal catalyzation. This remarkable adhesion improvement is due to the effect of formation of an intermixing layer at interface. As well-known, Cu and SiO_2 do not generally react at room temperature. Obviously, the noble metals lead to the intermixing acceleration, very likely by their catalytic effect.

These room-temperature/low-temperature processes achieve temperature reduction during microelectronic fabrication. In addition, it leads to formation of extremely thin (several nm thick) adhesion layer at the interface. The thickness reduction of adhesion layer can increase the area of the interconnection, and resistance of interconnection can be reduced. Therefore, these adhesion improvement processes at room temperature/low temperature is considered to be important for fabrication of a next-generation microelectronic device.

Acknowledgements

We acknowledge Professor Yoichi Nabetani and Professor Tsutomu Muranaka of the University of Yamanashi for providing ZnO films. A part of this work was financially supported by a Grant-in-Aid for Young Scientists (B) (25820122) from the Japan Society for Promotion of Science (JSPS) and Grant-in-Aid for Adaptable and Seamless Technology Transfer Program through target-driven R&D (AS262Z00926M) from the Japan Science and Technology Agency.

Author details

Mitsuhiro Watanabe[1]* and Eiichi Kondoh[2]

1 Department of Precision Machinery Engineering, College of Science and Technology, Nihon University, Funabashi, Japan

2 Interdisciplinary Graduate School, University of Yamanashi, Kofu, Japan

*Address all correspondence to: watanabe.mitsuhiro@nihon-u.ac.jp

IntechOpen

References

[1] Koike J, Wada M. Self-forming diffusion barrier layer in Cu-Mn alloy metallization. Applied Physics Letters. 2005;**87**:041911. DOI: 10.1063/1.1993759

[2] Neishi K, Aki S, Matsumoto K, Sato H, Itoh H, Hosaka S, et al. Formation of a manganese oxide barrier layer with thermal chemical vapor deposition for advanced large-scale intergrated interconnect structure. Applied Physics Letters. 2008;**93**:032106. DOI: 10.1036/1.2963984

[3] Yi SM, Jang KH, An JU, Hwang SS, Joo YC. The self-formatting barrier characteristics of Cu-Mg/SiO$_2$ and Cu-Ru/SiO$_2$ films for Cu interconnects. Microelectronics and Reliability. 2008;**48**:744-748. DOI: 10.1016/j.microrel.2007.12.005

[4] Russell SW, Rafalski SA, Spreitzer RL, Li J, Moinpour M, Moghadam F, et al. Enhanced adhesion of copper to dielectrics via titanium and chromium additions and sacrificial reactions. Thin Solid Films. 1995;**262**:154-167. DOI: 10.1016/0040-6090(94)05812-1

[5] Shepherd K, Niu C, Martini D, Kelber JA. Behavior of Cu$_{0.6}$Al$_{0.4}$ films at the SiO$_2$ interface. Applied Surface Science. 2000;**158**:1-10. DOI: 10.1016/S0169-4332(99)00525-5

[6] Schwalbe G, Baumann J, Kaufmann C, Gessner T, Koenigsmann H, Bartzsch A, et al. Comparative study of Cu and CuAl$_{0.3 wt.\%}$ films. Microelectronic Engineering. 2001;**55**:341-348. DOI: 10.1016/S0167-9317(00)00466-4

[7] Takagi H, Maeda R, Chung TR, Suga T. Low-temperature direct bonding of silicon and silicon dioxide by the surface activation method. Sensors and Actuators A. 1998;**70**:164-170. DOI: 10.1016/S0924-4247(98)00128-9

[8] Suga T, Takahashi Y, Takagi H, Gibbesch B, Elssner G. Structure of Al-Al and Al-Si$_3$N$_4$ interfaces bonded at room temperature by means of the surface activation method. Acta Metallurgica et Materialia. 1992;**40**:S133-S137. DOI: 10.1016/0956-7151(92)90272-G

[9] Takagi H, Kikuchi K, Maeda R, Chung TR, Suga T. Surface activated bonding of silicon wafers at room temperature. Applied Physics Letters. 1996;**68**:2222-2224. DOI: 10.1063/1.115865

[10] Howlader MMR, Zhang F. Void-free strong bonding of surface activated silicon wafers from room temperature to annealing at 600°C. Thin Solid Films. 2010;**519**:804-808. DOI: 10.1016/j.tsf.2010.08.144

[11] Kim TH, Howlader MMR, Itoh T, Suga T. Room temperature Cu-Cu direct bonding using surface activated bonding method. Journal of Vacuum Science and Technology A. 2003;**21**:449-453. DOI: 10.1116/1.1537716

[12] Sigetou A, Suga T. Modified diffusion bonding for both Cu and SiO$_2$ at 150°C. In: Proceedings of the 60th Electronic Components and Technology Conference. 2010. pp. 872-877. DOI: 10.1109/ECTC.2010.5490692

[13] Teraoka A, Watanabe M, Nabetani Y, Kondoh E. Room-temperature formation of ZnO-based adhesion layer for nanoprecision Cu/glass metallization. Japanese Journal of Applied Physics. 2013;**52**:05FB04. DOI: 10.7567/JJAP.52.05FB04

[14] Watanabe M, Teraoka A, Kondoh E. Room-temperature intermixing for adhesion enhancement of Cu/SiO$_2$ interface by adopting SiO$_2$ surface dope and noble metal catalyzation. Japanese Journal of Applied Physics.

2014;**53**:05GA02. DOI: 10.7567/
JJAP.53.05GA02

[15] Vietmeyer F, Seger B, Kamat
PV. Anchoring ZnO particles on
functionalized single wall carbon
nanotubes. Excited state interactions
and charge collection. Advanced
Materials. 2007;**19**:2935-2940. DOI:
10.1002/adma.200602773

[16] Matsumoto T, Mizuguchi T, Horii
T, Sano S, Muranaka T, Nabetani Y,
et al. Effects of Ga doping and substrate
temperature on electrical properties
of ZnO transparent conducting films
grown by plasma-assisted deposition.
Japanese Journal of Applied Physics.
2011;**50**:05FB13. DOI: 10.1143/
JJAP.50.05FB13

[17] Nakada T, Ohkubo Y, Kunioka
A. Effect of water vapor on the growth
of textured ZnO-based films for solar
cells by DC-magnetron sputtering.
Japanese Journal of Applied Physics.
1991;**30**:3344-3348. DOI: 10.1143/
JJAP.30.3344

[18] Kim YJ, Kim KW. Characteristics
of epitaxial ZnO films on
sapphier substrates deposited
using RF-magnetron sputtering.
Japanese Journal of Applied Physics.
1997;**36**:2277-2280. DOI: 10.1143/
JJAP.36.2277

[19] Okamura T, Seki Y, Nagakari S,
Okushi H. Preparation of n-ZnO/p-Si
heterojunction by sol-gel process.
Japanese Journal of Applied Physics.
1992;**31**:L762-L764. DOI: 10.1143/
JJAP.31.L762

[20] Kasuga M, Ogawa S. Electronic
properties of vaper-grown
heteroepitaxial ZnO film on sapphire.
Japanese Journal of Applied Physics.
1983;**22**:794-798. DOI: 10.1143/
JJAP.22.794

[21] Wenas WW, Ymada A, Takahashi
K. Electrical and optical properties of

boron-doped ZnO thin films for solar
cells grown by metalorganic chemical
vaper deposition. Journal of Applied
Physics. 1991;**70**:7119-7123. DOI:
10.1063/1.349794

[22] Viswanathan R, Gupta RB.
Formation of zinc oxide nanoparticles
in supercritical water. The Journal of
Supercritical Fluids. 2003;**27**:187-193.
DOI: 10.1016/S0896-8446(02)00236-X

[23] Konsoh E, Sasaki K, Nabetani
Y. Deposition of znic oxide thin films in
supercritical carbon dioxide solutions.
Applied Physics Express. 2008;**1**:061201.
DOI: 10.1143/APEX.1.061201

[24] Massalski TB, editor. Binary Alloy
Phase Diagrams. 2nd ed. OH: ASM
International; 1990. p. 1508

[25] Hamid M, Thir AA, Mazhar M,
Ahmad F, Molloy KC, Kociok-Kohn
G. Deposition and characterization of
ZnO thin films from a novel hexanuclear
zinc precursor. Inorganica Chimica
Acta. 2008;**361**:188-194. DOI: 10.1016/J.
ica.2007.07.013

[26] Zbels R, Muktepavela F,
Grigorjeva L, Tamanis E, Mishels-
Piesins M. Nanoindentation and
photoluminescence characterization
of ZnO thin films and single crystals.
Optical Materials. 2010;**32**:818-822.
DOI: 10.1016/j.optmat.2010.02.002

[27] Yoshiki H, Hashimoto K, Fujishima
A. Adhesion mechanism of electroless
copper film formed on ceramic
substrates using ZnO thin film as
an intermediate layer. Journal of the
Electrochemical Society. 1998;**145**:
1430-1434. DOI: 10.1149/1.1838500

[28] Sun RD, Tryk DA, Hashimoto K,
Fujishima A. Adheision of electroless
deposited Cu on ZnO-coated glass
substrates: The effect of the ZnO
surface morphology. Journal of the
Electrochemical Society. 1999;**146**:2117-
2122. DOI: 10.1149/1.1391901

[29] Hayashi C, Watanabe M, Kondoh E. Use of metal oxide as a Cu/glass adhesion promoting layer. Journal of the Surface Finishing Society of Japan. 2017;**68**:723-726. DOI: 10.4139/sfj.68.723

[30] Watanabe M, Koike K, Kondoh E. Improvement in adhesion of Cu/glass stacks using ZnO thin films deposited by chemical solution methods and its formation conditions. Journal of the Surface Finishing Society of Japan. 2015;**66**:534-539. DOI: 10.4139/sfj.66.534

[31] Watanabe M, Tamekuni S, Kondoh E. Formation of zinc oxide thin film using supercritical fluids and its application in fabricating a reliable Cu/glass stack. Microelectronic Engineering. 2015;**141**:184-187. DOI: 10.1016/j.mee.2015.03.031

[32] Bunting EN. Phase equiliblia in the system SiO_2-ZnO. Bureau of Standards Journal of Research. 1930;**4**:131-136. DOI: 10.1111/j.1151-2916.1930.tb16797.x

Chapter 5

A Review: Solder Joint Cracks at Sn-Bi58 Solder ACFs Joints

Shuye Zhang, Tiesong Lin, Peng He and Kyung-Wook Paik

Abstract

In this chapter, solder joint cracks at Sn-Bi58 solder ACF joints were investigated in conventional thermal compression bonding and ultrasonic bonding. It was found that resin storage modulus is the crucial for solder joint morphology regardless of bonding pressures. At high temperature, polymer resin tends to rebound above Tg and break the molten solder morphology. We proposed two useful methods to keep off solder joints cracks during bonding process. One is to remain bonding pressure until room temperature, the other is to use fillers to increase resin thermal mechanical property. The thermal cycling reliability was significantly enhanced when solder joint morphology was modified using 10 wt% 0.2 μm SiO_2 fillers in acrylic based Sn-Bi58 solder ACF joints.

Keywords: solder cracks, ACF assembly, flex-on-board assembly, high reliability

1. Introduction

In 2013, Google Company had launched Google Glass, which the first-generation of wearable electronics in the history of humans [1]. Apart from the limited packaging sizes, high-density packaging technologies are demand for chips, passive components, and printed circuit boards. Several functions such as cameras, global positioning system (GPS), wireless communication, touch screen, FM radio, Audio, are also featured in Google Glass [2]. Generally, socket-type connectors have been used to connect between a flexible printed circuit (FPC) module and the main board of Google Glass [3], on the purpose of electrical interconnect. Flex-on-board (FOB) is one type of flip-chip technologies, to assembly printed circuits board (PCB) and FPC using anisotropic conductive films (ACFs) [4].

Aiming at replacing the socket-connectors, FOB assembly is attracting more and more attention, due to a lower thickness (about 50 μm) and a higher fine-pitch capability (under 100 μm) [5]. So Google started to use FOB in mother board assembly to partly take place of connectors, as shown in **Figure 1**. ACFs are usually to be as the interconnection materials to assembly FOB, consisting of thermosetting polymer adhesive matrix and conductive balls [6]. Adhesives will be cured by temperature and functional group will be cross-linked, resulting into the mechanical connection to PCB board and metal pad surfaces [7]. Current flows through an ACF joint formed by a physical contact between electrodes and conductive balls (such as Au/Ni metal balls or Au/Ni coated polymer balls) [8].

Figure 2 shows an Au/Ni metal ACF joint and a Sn based metallurgical ACF joint in a cross-section view using a scanning electron microscope. Compared with

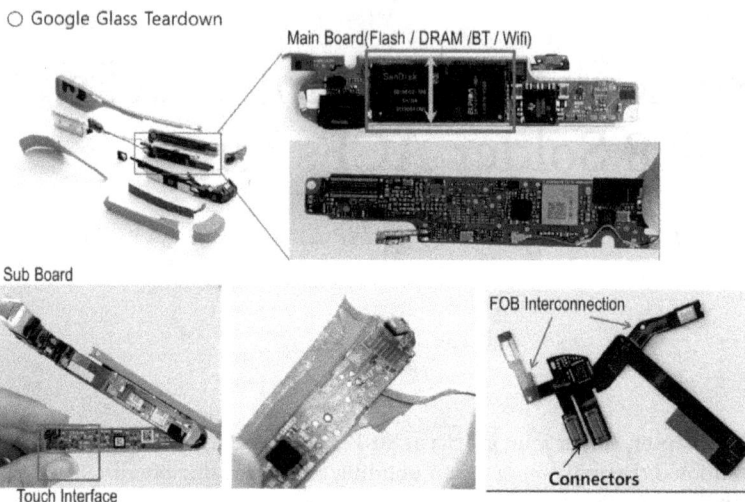

Figure 1.
Google glass teardown and FOB interconnection.

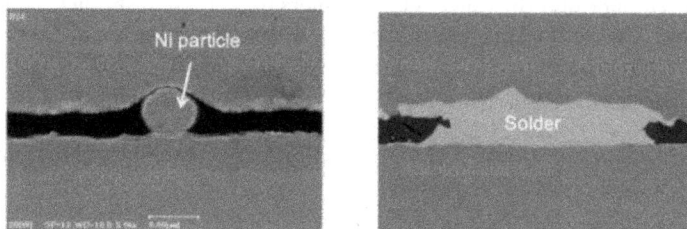

Figure 2.
A comparison of the conventional Ni and Sn solder metallurgical anisotropic conductive films (ACFs) joints.

a metal ACF joint, high contact resistance, poor power handling capacity and reliability can be improved by using solder metallurgical ACF joint, due to wide electrical paths and stable metallurgical interconnection [9, 10]. In order to remove solder oxide layer and improve solder wettability, two methods, a thermal compression (TC) bonding combining a flux material [11] and an ultrasonic (US) bonding without flux materials [12], are used. According to previous results, the heating rates were raised rapidly to 400°C/s and the temperature of solder ACF joints reached above 250°C under US vibration. By adjusting various ultrasonic amplitudes of vibration (from 4 to 13 μm), the ACF temperature could be precisely controlled from 70 to 250°C.

A perfect solder ACF joint morphology containing a Sn–3Ag–0.5Cu (SAC305) alloy has been optimized by using lower viscosity, faster curing speed, higher resin property based cationic epoxies with high elastic modulus on a 250°C bonding temperature for FOB assembly [13]. Low viscosity helps resin flow during bonding process [14] and faster curing speed indicates higher cross-linking density and mechanical property of polymer resins [15]. Compared with acrylic resin, imidazole resin and multifunctional epoxy enhanced imidazole resin, cationic epoxy resin has the highest elastic modulus when it is fully cured, therefore, few solder joint cracks are taken place at ACF joints after FOB assembly. Not only resin property is a basic issue, bonding time also plays an important role in solder joint morphologies, especially for cracks [16]. Since micron sized solder ACF joint is so tiny that Sn elements quickly diffused into Au/Ni metal surface to form brittle intermetallic compound (IMC), cracks were formed by the residual Bi phase and the newly formed IMC phases. As a result, 10 seconds bonding time was optimized to avoid over-diffusion behavior of Sn elements at micron solder ACF joints.

At optimized 10 seconds bonding time, depending on their solder melting temperatures, SAC305 (221°C) and Sn-58Bi (139°C) solder ACFs are bonded at 250 and 200°C joint temperatures, respectively. For those resins with high elastic modulus (such as imidazole resin, multifunctional epoxy enhanced imidazole resin, and cationic epoxy resin), solder cracks were rarely found at ACF joints after TC assembly in FOB interconnection. On the contrary, for these resins with low elastic modulus (such as acrylic resin), solder joint crack was a critical issue after FOB assembly [17]. Although bonding temperature was suggested not higher than 220°C for acrylic ACF resins to avoid thermal decomposition and solder joint cracks, solder joint cracks were even observed at low modulus based acrylic resin joints at 200°C bonding temperatures.

Although polymer rebound of the cured acrylic resin had been measured as approximate 1–3% dimension change of polymer resin, when the bonding pressure was released at 200°C bonding temperature [17]. In this chapter, we aimed at finding out the obvious inner factor of polymer resins to determine Sn-58Bi solder cracks after FOB assembly at acrylic ACF joints, rather than a perfect solder joint morphology using other ACF resins. After that, two available throughout methods were discussed to increase acrylic resin elastic modulus to solve solder joint crack after FOB assembly. Moreover, the consequent solder joint morphologies were observed, compared and analyzed. The significance of this research is to guide ultra-low elastic modulus ACF resin assembly to form reliable solder joints for low melting solder materials and electronic device packaging.

2. Experiments

2.1 Test vehicles and materials

Test vehicles was shown in **Figure 3**. FR-4 printed circuit board (PCB) was 1-mm-thick and flexible printed circuit (FPC) board was made by polyimide with 50-μm-thick, and 500-μm-pitch Cu patterns with electroless nickel immersion gold (ENIG) finish were plated on test vehicles.

Three kinds of polymer resins were compared for the ACFs, acrylic resin, imidazole resin and cationic resin. These products were all bought from H&S company in South Korea. About 5 wt% 8-μm-diameter Ni particles, 0.2 μm silica fillers,

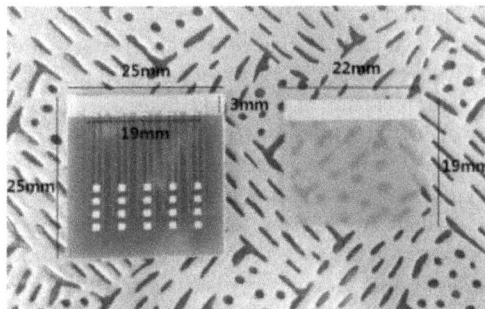

Figure 3.
500-μm-pitch printed circuit board (PCB) and flexible printed circuit (FPC) board.

Solder ACFs	Weight percentage				Calculated volume percentage			
	Sn-58Bi solder (8.56 g/cm³)	Polymer resin (1.25 g/cm³)	Ni particle (8.9 g/cm³)	Silica filler (2.65 g/cm³)	Sn-58Bi solder (8.56 g/cm³)	Polymer resin (1.25 g/cm³)	Ni particles (8.9 g/cm³)	Silica filler (2.65 g/cm³)
ACF 1	30%	1	5%	0%	6.25%	92.75%	1%	0%
ACF 2	30%	1	5%	5%	6.1%	89.7%	0.97%	3.23%
ACF 3	30%	1	5%	10%	5.9%	86.9%	0.93%	6.27%

Table 1.
The specification of solder ACFs.

30 wt% 25–32 μm diameters Sn-58Bi particles, and 2 wt% flux material were added in the pure resins and then proceeded the film coating process. After that, 50-film-thickness anisotropic conductive film was achieved. **Table 1** gives the specifications of the added materials, such as weight percentages of the pure polymer resins and the calculated volume percentages of the total ACFs materials with additives.

Molten solder joint was like a water above its melting temperature and squeezed by bonding pressure, resulting into uniform solder joint morphology after bonding [18–20]. Thus, Ni particles were used to obtain uniform joint gap size and control solder joint morphology. Before coating process of ACF, polymer matrixes and conductive particles were stated in a solution. Silica fillers were gradually added as a function of weight percentage by hands [21–23]. Since the solution viscosity was be very high when adding the nano-size solid silica fillers, plenty of toluene would be added to decrease the solution viscosity and make sufficient dissolution between nano sized silica fillers and polymer matrixes. In addition, considering the agglomeration of nano size silica fillers, a magnetic stirring apparatus was carried out at 30° C for 0.5 hour and then a rolling mechanical stirring was performed at room temperature for 12 hours. As a result, a sufficient dissolution between nano size silica filler and polymer resin solution was achieved. **Figure 7** shows the fresh ACFs were put on the PCB ENIG metal electrodes before bonding process.

2.2 Bonding methods

2.2.1 Thermo-compression bonding

According to previous results [17], the bonding parameters (peak temperature, time, and pressure) were optimized as 200°C 10 seconds and 1 MPa. **Figure 4** gives a heat conduction mechanism from a hot bar to ACF joint on a FOB application. Thermal setting adhesives were cured due to the heat conduction from the high temperature of hot bar. **Figure 5** shows a clear temperature profile according to previous result, where temperature was quickly up to Sn-58Bi solder melting point (139°C) and then gradually reached the 200°C peak temperature at 4 seconds. Afterwards, solder ACF joint temperature was remained at 200°C from the 4th to 10th second, and then solder ACF joint entered cooling procedure without any pressure protect.

However, it showed solder joint temperature reached 139°C in the cooling process until 11th second, which means from 10th to 11th second solder ACF joint freely cooled and was not protected with pressure. It is well known that molten solder joint is very weak and its morphology is easy to be destroyed by polymer property, for example, polymer will be rebound on the moment of TC bonding finish and hot bar releasing from test vehicles [24]. Therefore, the mismatch between resin property and molten Sn-58Bi solder property under the cooling

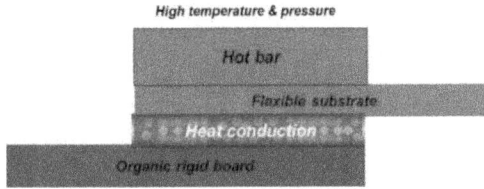

Figure 4.
Heat conduction mechanism in a TC bonding.

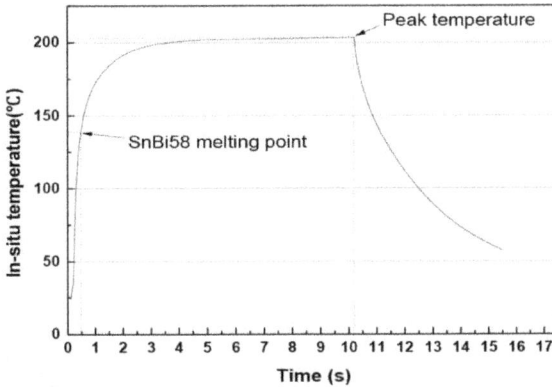

Figure 5.
The in-situ temperature profile of solder ACFs joint by a TC bonding.

process should be investigated. At here, in-situ temperature is precisely measured by a K-type thermocouple every 0.1 seconds [25].

2.2.2 Ultrasonic bonding

Figure 6 shows another mechanism of local heat generation by ultrasonic vibration at room temperature. Unlike TC bonding, US bonding was applied at room temperature and bonding pressure can be controlled as a function of time, which means the bonding pressure can be maintained through the whole bonding procedure until ACF joint cooling to room temperature [25]. In this way, ACF adhesives and solder joint were protected under bonding pressure during cooling process, to reduce the influence of heated resin to molten solder joints. During the US bonding, polymer resin can also be cured by the spontaneous ultrasonic vibration at room temperature environment. Compared with TC bonding, the joint temperature was slowly

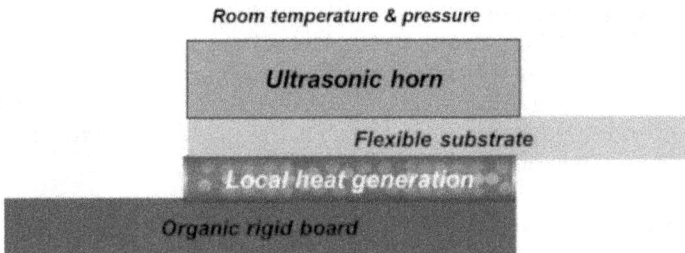

Figure 6.
Mechanism of local heat generation by ultrasonic vibration at room temperature.

Effects of Lift-up hot bar during cooling process

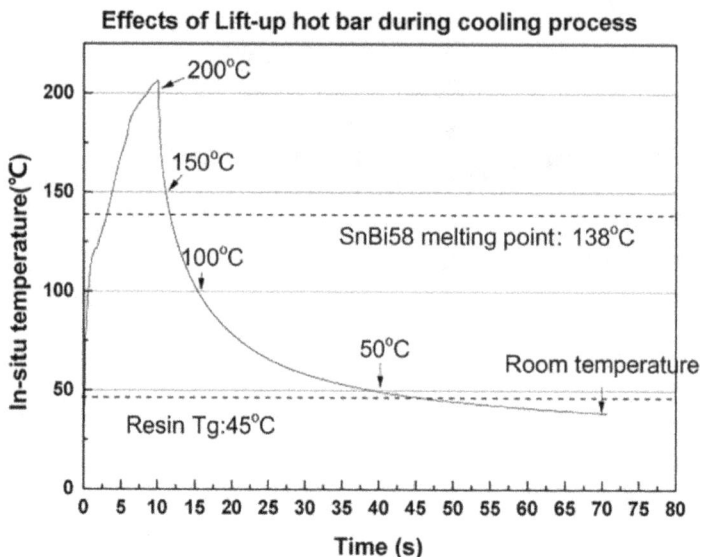

The in-situ temperature and the designed lift-up time of bonding pressure in US bonding.

increased up to 200°C, however, resin would be fully cured and solder metallurgical joint could be obtained as well [26]. **Figure 7** shows the in-situ ACFs joint temperature and the designed lift-up time of bonding pressure in US bonding. Referring to previous results, polymer rebound amount will be decreased as temperature decreased [17]. At here, we focus whether it is possible to prevent the solder joint damage at a lower temperature releasing bonding pressure during the cooling process.

2.3 Differential scanning calorimetry (DSC)

Many studies have been reported the curing degree of acrylic ACF joints by TC or US bonding using DSC [14, 17, 26, 27]. It has been convinced that acrylic resin can be fully cured at 200°C 10 seconds condition and provide the best thermo-mechanical property after its full cure [28]. Considering there is an exothermic peak from 75 to 150°C caused by acrylic resin cure and a sharp endothermic peak from 135 to 142°C caused by Sn-58Bi solder melting, there might be interplay by two materials and unclear to demonstrate the Sn-58Bi solder melting in the temperature heating-up period. On the other hand, it is important to know the solidification temperature of Sn-58Bi alloy during the temperature cooling-down period. Because mechanical protection of Sn-58Bi alloy will be established when alloy is solidified [29, 30].

Therefore, only pure Sn-58Bi alloy was tested by DSC to obtain the melting and solidification temperature, respectively. In details, Sn-58Bi solder alloy was put by 20 mg weight. The heating rate was 20°C/minutes from 30 to 300°C, afterwards, the temperature was along with furnace cooling to room temperature at a nitrogen environment. The curing behavior of acrylic resin used in this study and curing degree after bonding process measured by a Fourier transform infrared spectroscopy can be found at here [14, 17, 26, 27].

2.4 Thermomechanical analysis

A thermomechanical analyzer (DMA) was used to measure the storage modulus of the cured polymer resin as a function of temperature. The resin film was prepared by the thickness of 50 μm, the length of 10 mm, and the width of 2.5 mm, and it was cured at the oven environment of 150°C for 3 hours. In DMA environment, resin film was tested under a 0.1 Hz load with a 10 mN dynamic stress, and the static tensile force was prepared by 50 mN. In addition, the DMA environment was heated by 5°C/ minutes until 200°C from room temperature. The elastic property of polymer resin in this study was considered in z direction, because the polymer rebound tool place vertically, resulting into the ACF dimension change and solder joint cracks. As Eq. (1) was shown in the following, elastic strain is mathematically determined by the ratio of length change (ΔL) over the initial ACF length (L). The larger changed length of polymer resin by DMA force indicates the larger polymer rebound when bonding tool was disappeared. In addition, the coefficient of thermal expansion (CTE) of adhesives were also measured by a 50 mN tensile force.

$$Strain = (\Delta L)/L \qquad (1)$$

2.5 Joint resistance and morphologies

After bonding process, ACFs joint was formed and its resistance was detected by a contact method, which is called 4-point-probe kelvin method. Referring to Ohm's Law, we have learnt that resistance is mathematically named by the ratio of electrical voltage above its corresponding current. As a constant current was applied through all over the circuit, the resistance of the specific part is easy to know unless the voltage is precisely measured [31].

Actually, a delta mode in KI 6220 nano-voltmeter has been widely used to measure the microscale ohms from the American Keithley company. The constant current was designed and perfectly applied in Cu lines at PCB and FPC substrates, as a result, solder ACFs joints were performed by current, voltage drop by can be measured by KI 6220 m, which was shown as the grown overlapped part in **Figure 8**. For accurately measuring the contact resistance of solder ACF joint, one measurement is repeated by 10 times. In details, Cu pad areas was 0.3 mm² and 40 channels were designed in PCB test vehicles.

A scanning electron microscope (SEM) in this paper was utilized to apparently observe the changing of solder ACFs joint morphologies before and after sever cracks. In order to obviously compare Sn and Bi elements in solder joint morphologies, a backscattered electron mode was carried out in SEM environment. For the typical solder morphologies, such as solder joint heights, shapes, and cracks, we performed at SEM images and evaluated them at certain conditions.

2.6 Reliability evaluation

To characterize cracks at solder joint morphologies on the effects of electrical performance, a thermal cycling reliability from −45 to 125°C was tested until 1000 cycles. TC bonded Sn-Bi58 solder ACFs joints with 0, 5, and 10 wt% of 0.2 μm

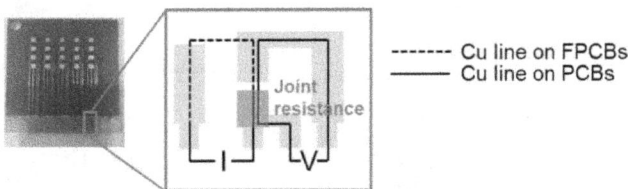

Figure 8.
Electrical design for a four-point-probe measurement.

SiO_2 fillers addition were compared in this thermal cycling test. What is more, the dwell time of -45°C was remained for 15 minutes and it rapidly increased to 125°C for 15 minutes dwelling time. Joint contact resistances in this thermal cycling test were designed to be recorded every 200 cycles. Until 1000 cycles, joint morphologies and each failure mode of Sn-Bi58 solder ACF joints with or without silica modified were compared through the observation of SEM images.

3. Results and discussion

3.1 Effects of resin modulus on solder joint cracks

Figure 9 shows the thermal mechanical properties of typical adhesives films as a function of temperatures. Compared with conventional epoxies by imidazole and cationic curing types with higher Tg (over 100°C), acrylic adhesives showed lower modulus, because it was a low Tg material (45°C). Especially at 200°C when hot-bar releasing bonded samples at the end of TC bonding for Sn-58Bi solder ACFs applications, even cured acrylic adhesives showed 0.4 Mpa storage modulus, and cationic epoxy and imidazole epoxy showed 7.1 and 5.7 Mpa, respectively.

Figure 10 shows Sn-58Bi solder ACFs joints cracks using this low Tg and low modulus acrylic adhesives regardless of bonding pressures from 1 to 3 Mpa. However, there are no solder joint cracks when using other higher storage modulus adhesive materials under the same bonding pressures.

In terms of solder joints cracks, there are two assumptions on it: one is CTE mismatch between shrink adhesives and solders during cooling process, the 2nd is higher elasticity of adhesives than molten solders when bonding pressure is removed, and then liquid solder joints are cracked under high temperature.

Figure 11 shows endothermic and exothermic behaviors of Sn-58Bi solder materials in a DSC analysis. In a heating period, Sn-58Bi solder melts at 139°C and stay at a liquid state until cooling to 125°C, and Sn-58Bi solders were totally solid below 90°C. As a result, adhesives shrinkage or rebound is able to cause liquid Sn-58Bi solder joints cracks only above 90°C, because solid Sn-58Bi joint is with 30.9 Gpa young's modulus [32] which cannot be cracked. **Figure 12** gives the thermal expanded behaviors of three typical adhesives. The shrinkage percentages of three adhesives films were summarized in **Table 2** in cooling process. 90°C is lower than the imidazole epoxy's Tg and cation epoxy's Tg, as a result, there was a huge shrinked amount when cooling process was lower than glass transition temperature

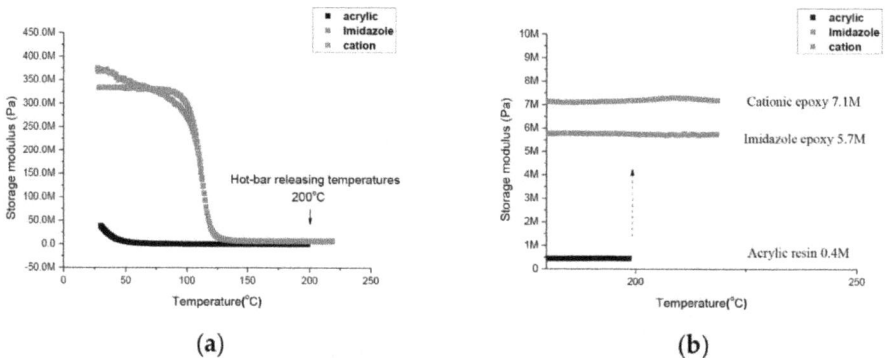

(a)

(b)

Figure 9.
*Storage modulus of typical adhesives films as function of temperature. (**a**) 30–250°C (**b**) specific region at 200–250°C.*

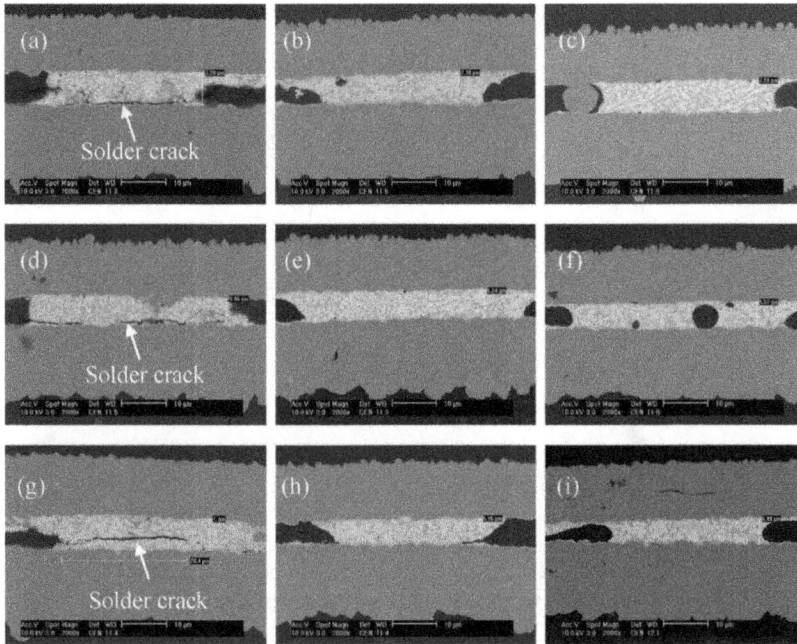

Figure 10.
*Solder joint morphologies of Sn-Bi58 ACFs joints by various bonding pressures. (**a, d, g**) Acrylic resins bonded by 1, 2 and 3 MPa, respectively; (**b, e, h**) imidazole epoxy bonded by 1, 2 and 3 MPa, respectively; (**c, f, i**) cationic epoxy bonded by 1, 2 and 3 MPa, respectively.*

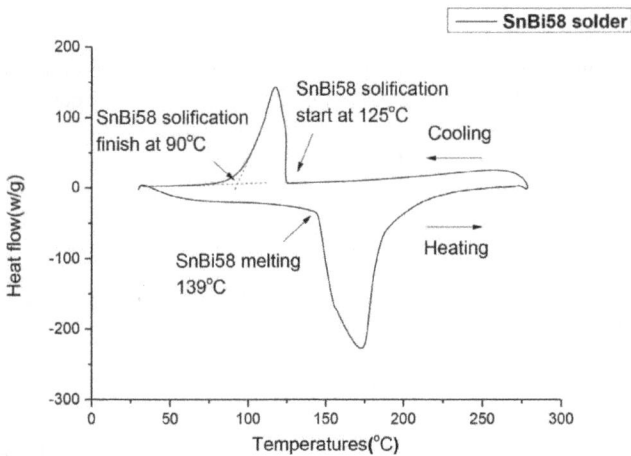

Figure 11.
DSC behaviors of Sn-58Bi solder during heating and cooling process.

at two epoxies. −11.2, −13.2 and −5.2% dimension shrinkage were shown for acrylic resin, imidazole epoxy and cation epoxy in the cooling region from 200 to 90°C. However, good solder joint morphologies were found at imidazole epoxy based Sn-58Bi solder ACFs joints in **Figure 10**. As a result, solder ACFs joint cracks were not related with compressive stress by adhesives shrinkages in the cooling process, but related with resin elasticity, especially for lower modulus acrylic adhesives.

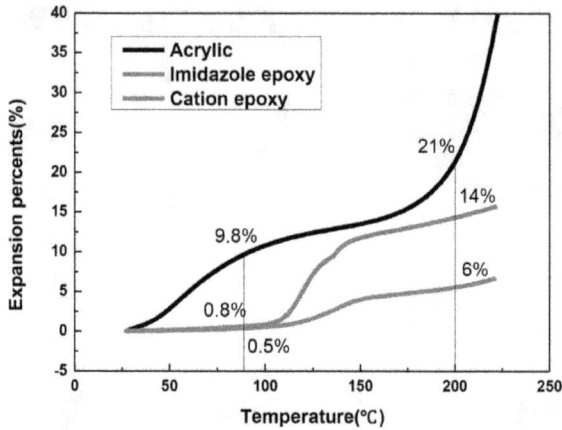

Figure 12.
Thermal expansion properties of typical adhesives films.

Adhesive modulus is divided by storage and loss modulus in viscoelastic materials [33]. Storage modulus is to measure the stored energy and represent the elastic property, and loss modulus is to measure the energy dissipated as heat and represent the viscous property [34]. In this study, solder joint cracks were due to adhesive rebound, which is elastic properties, and not due to heat dissipation and adhesive viscos property, so storage modulus is used.

Figure 13 illustrates the strain changes of acrylic based Sn-58Bi solder ACFs with respect to the 50 mN tensile sinusoidal load with 10 mN amplitude and 0.1 Hz frequency in DMA analysis as a function of temperatures. In details, strain changes in **Figure 13** were consisted of two parts, one part is due to thermal expansion and the other is elastic strain due to plastic deformation. In this study, the plastic deformation, which is the dimension recover of deformed polymer when mechanical loading disappeared, is used to estimate adhesive rebound. According to the following equation, storage modulus is the ratio of applied tensile stress to elastic strain. In other words, adhesive rebound amounts are reversely proportional to its storage modulus in the following Eq. (2).

$$Storage\ modulus = Stress/(Elastic\ strain) \qquad (2)$$

Because Tg of acrylic resin is 45°C, adhesive showed a relatively higher modulus at room temperatures below 45°C and lower modulus above 45°C. As a result, the elastic strain is smaller below 45°C and larger at above 45°C. In addition, 30 wt% Sn-58Bi solder particles melting behavior at 139°C will enlarge the measured sample dimension in **Figure 13**. **Figure 14** shows storage modulus of acrylic adhesives were increased during cooling process and **Table 3** summarizes specific storage modulus of acrylic adhesive during cooling process.

Dimension change	Acrylic	Imidazole epoxy	Cation epoxy
200°C	21%	14%	6%
90°C	9.8%	0.8%	0.8
200 Cooling to 90°C	−11.2%	−13.2%	−5.2%

Table 2.
Shrinkage percentages of typical adhesives films during cooling.

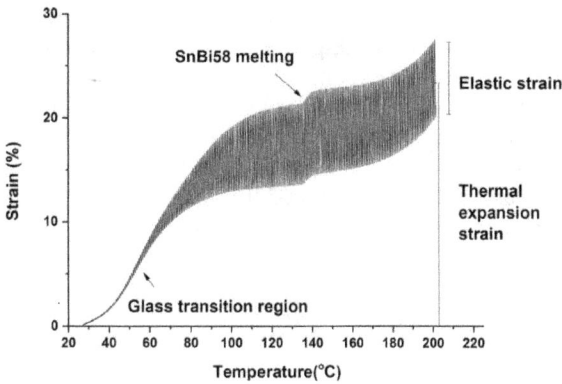

Figure 13.
Strain of acrylic based Sn-58Bi solder ACFs in respect to a sinusoidal load applied in the DMA test as a function of temperatures.

3.2 Effects of US bonding on enhancing resin modulus

Figure 15 shows the effects of delaying ultrasonic horn lift-up time on the acrylic based Sn-58Bi solder ACFs joint morphologies during an US bonding method. A crack-free Sn-58Bi solder ACFs joint can be successfully obtained by maintaining pressure below 45°C the acrylic adhesives Tg during a cooling process, because high storage modulus was established at low temperatures. Because 90°C is the complete point of Sn-58Bi solder joint solidification in **Figure 11**, there was a still solder joint crack when removing bonding pressures at 100°C in **Figure 15**. However, over 30 seconds remaining pressure time is too long for assembly.

3.3 Effects of silica filler on enhancing resin modulus

The faster approach of removing the solder joint crack is to increase resin storage modulus over Tg by adding silica fillers into acrylic polymer resins.

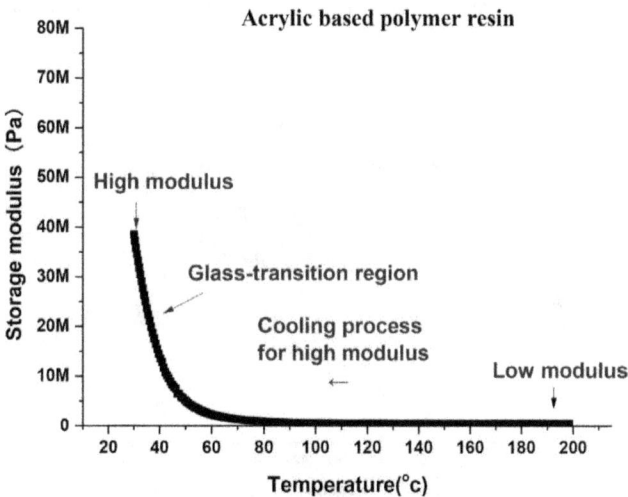

Figure 14.
Increased storage modulus of acrylic solder ACFs during cooling process.

Figure 16 shows the strain amount decrease of acrylic based Sn-58Bi solder ACFs as a function of temperature by adding 0, 5 and 10 wt% of 0.2 μm silica fillers. Both the amount of the thermal expansion and the elastic strain of polymer resin were reduced by adding silica filler. It was known that reducing the thermal expansion strain and the elastic strain as a function of temperatures are good for the joint reliability in T/C test and the good solder joint formation, respectively [35]. Storage modulus of acrylic based Sn-58Bi solder ACFs with addition of 0, 5, and 10 wt% 0.2 μm silica was increased from 0.4 to 0.9 and 1.3 Mpa in **Figure 17**, respectively.

Temperature	200°C	150°C	100°C	50°C	30°C
Modulus	0.4 MPa	0.44 MPa	0.53 MPa	5.2 MPa	34 MPa

Table 3.
Storage modulus of acrylic adhesive during cooling process.

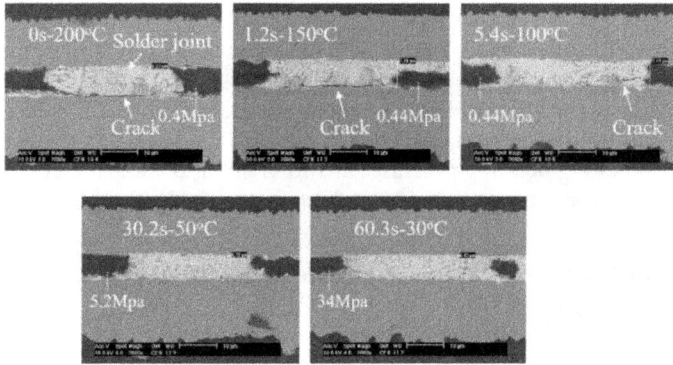

Figure 15.
Effects of delaying ultrasonic horn lift-up times (0, 1.2, 5.4, 30.2, and 60.3 seconds) on the Sn-58Bi solder ACFs joint morphologies.

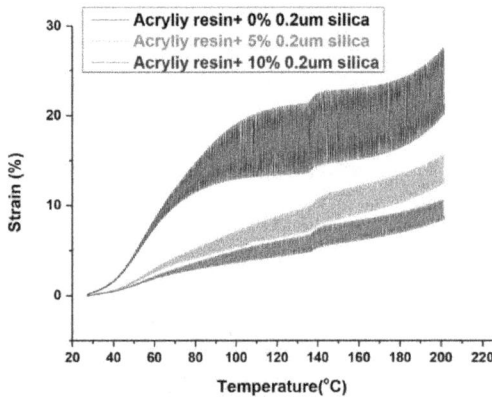

Figure 16.
Strain of acrylic based Sn-58Bi solder ACFs with added silica fillers as a function of temperatures.

3.4 T/C reliability

Figure 18 were listing the solder joint morphologies by adding 0, 5 and 10 wt% 0.2 μm sized silica filler of acrylic based Sn-58Bi solder ACFs joints before and after T/C reliability for 1000 cycles. Referring to the enhancement of elastic modulus by adding several silica fillers in **Figure 17(b)**, solder joint cracks were completely removed for acrylic based Sn-58Bi solder ACF joints at 200°C TC bonding condition with several bonding pressures.

For acrylic based Sn-58Bi solder ACFs without added silica fillers, total joint failure occurred at the interface between Sn-58Bi solder joints and Cu metal electrode after 1000 cycles T/C reliability test, because initial solder joint cracks have already existed before the T/C reliability test. Although elastic modulus was increased double by adding 5 wt% 0.2 μm silica fillers, a small solder crack still remained and propagated at Sn-58Bi solder joints resulting in unstable joint contact resistance during T/C reliability. Excellent solder joint morphology was obtained after the 1000 cycles T/C reliability, because initial solder joint crack was perfectly

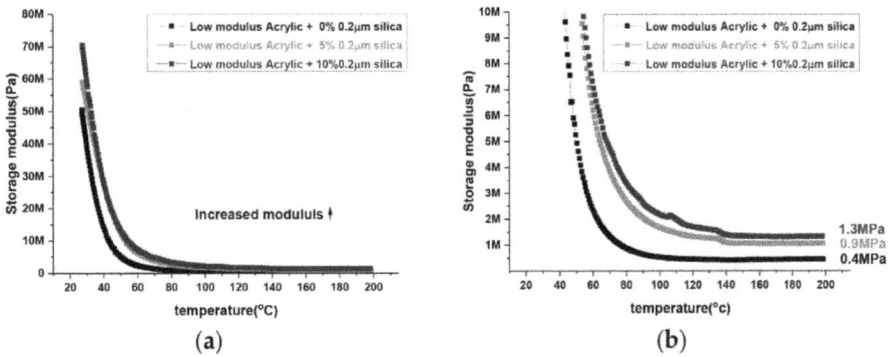

Figure 17.
Elastic modulus of acrylic based Sn-58Bi solder ACFs by adding 0, 5, and 10 wt% 0.2 μm silica fillers at (**a**) 0–80 MPa and (**b**) 0–10 MPa ranges.

Figure 18.
Solder joint morphologies of acrylic based Sn-58Bi solder ACFs added by 0, 5, and 10 wt% 0.2 μm silica filler before and after 1000 cycles reliability test.

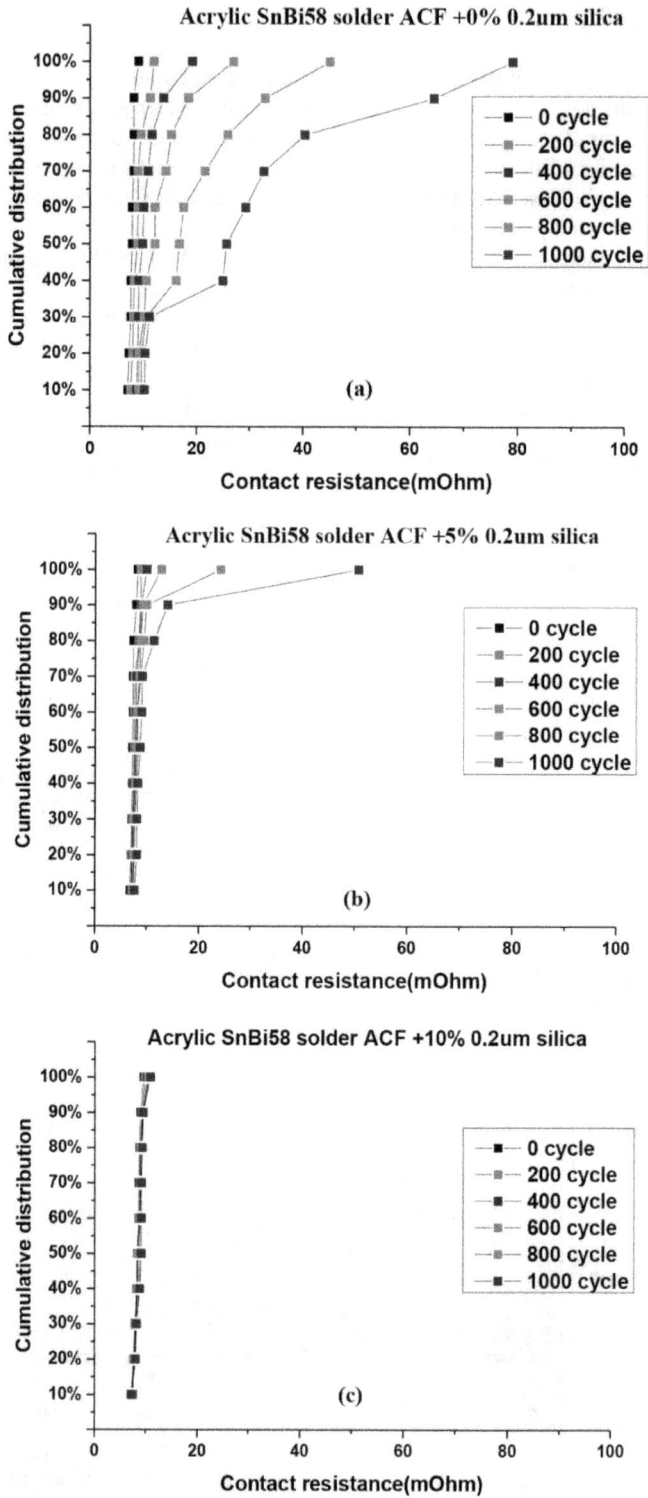

Figure 19.
Solder joint contact resistances of acrylic based Sn-58Bi solder ACFs up to 1000 cycles T/C reliability as a function of (a) 0, (b) 5, and (c) 10 wt% 0.2 μm silica fillers addition.

removed at Sn-58Bi solder ACF joints by adding 10 wt% 0.2 μm silica fillers. **Figure 19** shows the 1000 cycles T/C reliability results in terms of contact resistance, and stable joint contact resistance was achieved by adding 10 wt% 0.2 μm silica fillers due to crack-free solder joint.

4. Conclusion(s)

In this paper, we investigated the resin properties and bonding parameters on the solder joint morphologies of Sn-Bi58 ACF joints. As a result, we found storage modulus of resin adhesives was the determined factor for solder joint cracks and regardless of bonding pressures. We thought cracks at solder joints happened, probably due to the high elasticity of polymer resin. Apart from that, two suggestions were listed to solve the solder joint cracks by increasing the resin storage modulus. The 1st one was to remain the hot-bar until cooling to its Tg, but this method will excessively consume solder and lead to brittle IMC at interfaces in TC bonding, while it is ok to US bonding. The other method was to add silica fillers in polymer resin to increase its thermos-mechanical property and reduce the polymer rebound when bonding process was finished. For ultrasonic bonding, storage modulus above 5 MPa of was at least needed to prevent solder joint cracks. On the contrast, 70 seconds for maintaining bonding pressure was too long. More than 1.38 MPa storage modulus at 200°C was needed for a crack-free Sn-58Bi solder joint morphology for a conventional TC bonding.

Acknowledgements

The authors thank National Natural Science Foundation of China (Grant 51805115) for research funding support.

Conflict of interest

The authors declare no conflict of interest.

Author details

Shuye Zhang[1*], Tiesong Lin[1], Peng He[1*] and Kyung-Wook Paik[2]

1 State Key Laboratory of Advanced Welding and Joining, Harbin Institute of Technology, Harbin, China

2 Department of Materials Science and Engineering, KAIST, Daejeon, South Korea

*Address all correspondence to: syzhang@hit.edu.cn; hithepeng@hit.edu.cn

IntechOpen

References

[1] Stoppa M, Chiolerio A. Wearable electronics and smart textiles: A critical review. Sensors. 2014;**14**(7):11957-11992

[2] Ladner RE. Communication technologies for people with sensory disabilities. Proceedings of the IEEE. 2012;**100**(4):957-973

[3] Zhang S, Kim SH, Kim TW, et al. A study on the solder ball size and content effects of solder ACFs for flex-on-board assembly applications using ultrasonic bonding. IEEE Transactions on Components Packaging & Manufacturing Technology. 2015;**5**(1): 9-14

[4] Sjoberg J, Geiger DA, Shangguan D. Process development and reliability evaluation for inline Package-on-Package (pop) assembly. Electronic Components and Technology Conference, ECTC 2008. 58th. IEEE; 2008:2005-2010

[5] Zhang S et al. Mechanism of solder joint cracks in anisotropic conductive films bonding and solutions: Delaying hot-Bar lift-up time and adding silica fillers. Metals. 2018;**8**(1):42

[6] Jung KS et al. Anisotropic conductive film forming composition. U.S. Patent No. 7,700,007. Ap. 20, 2010

[7] Kim S-C, Kim Y-H. Flip chip bonding with anisotropic conductive film (ACF) and nonconductive adhesive (NCA). Current Applied Physics. 2013;**13**:S14-S25

[8] Liu J, Salmela O, Sarkka J, et al. Reliability of microtechnology: Interconnects, devices and systems. Springer Science & Business Media. 2011

[9] Zhang S, Park JH, Paik KW. Joint morphologies and failure mechanisms of anisotropic conductive films (ACFs) during a power handling capability test

for flex-on-board applications. IEEE Transactions on Components Packaging & Manufacturing Technology. 2016;**99**: 1-7

[10] Zhang S, Paik KW. A study on the failure mechanism and enhanced reliability of Sn58Bi solder anisotropic conductive film joints in a pressure cooker test due to polymer viscoelastic properties and hydroswelling. IEEE Transactions on Components Packaging & Manufacturing Technology. 2016; **6**(2):216-223

[11] Kim SH, Choi Y, Kim Y, Paik KW. Flux function added solder anisotropic conductive films (ACFs) for high power and fine pitch assemblies. In: Proceedings of the Electronic Components and Technology Conference (ECTC), 2013 IEEE, Las Vegas, NV, USA. May 28–31, 2013. pp. 1713-1716

[12] Lee K, Saarinen IJ, Pykari L, Paik KW. High power and high reliability flex-on-board assembly using solder anisotropic conductive films combined with ultrasonic bonding technique. IEEE Transactions on Components, Packaging and Manufacturing Technology. 2011;**1**: 1901-1907

[13] Zhang S, Yang M, Wu Y, et al. A study on the optimization of anisotropic conductive films for Sn-3Ag–0.5Cu-based flex-on-board application at a 250°C bonding temperature. IEEE Transactions on Components Packaging and Manufacturing Technology. 2018;**8** (3):383-391

[14] Kim YS, Lee K, Paik KW. Effects of ACF bonding parameters on ACF joint characteristics for high-speed bonding using ultrasonic bonding method. IEEE Transactions on Components Packaging & Manufacturing Technology. 2013; **3**(1):177-182

[15] Keller A et al. Fast-curing epoxy polymers with silica nanoparticles: Properties and rheo-kinetic modelling. Journal of Materials Science. 2016;**51**(1): 236-251

[16] Zhang S et al. A study on the bonding conditions and nonconductive filler contents on cationic epoxy-based Sn-58Bi solder ACFs joints for reliable flex-on-board applications. IEEE Transactions on Components Packaging & Manufacturing Technology. 2017;**99**: 1-8

[17] Zhang S et al. Effects of acrylic adhesives property and optimized bonding parameters on Sn58Bi solder joint morphology for flex-on-board assembly. Microelectronics Reliability. 2017;**78**:181-189

[18] Kim Y-S, Zhang S, Paik K-W. Highly reliable solder ACFs FOB (flex-on-board) interconnection using ultrasonic bonding. Journal of the Microelectronics and Packaging Society. 2015;**22**(1):35-41

[19] Zeng K, Tu KN. Six cases of reliability study of Pb-free solder joints in electronic packaging technology. Materials Science and Engineering: R: Reports. 2002;**38**(2):55-105

[20] LoVasco F, Oien M A. Process for controlling solder joint geometry when surface mounting a leadless integrated circuit package on a substrate. U.S. Patent 4,878,611 [P]. Nov 7, 1989

[21] Teh PL, Mariatti M, Akil HM, et al. The properties of epoxy resin coated silica fillers composites. Materials Letters. 2007;**61**(11–12):2156-2158

[22] Sangermano M, Malucelli G, Amerio E, et al. Photopolymerization of epoxy coatings containing silica nanoparticles. Progress in Organic Coatings. 2005; **54**(2):134-138

[23] Douce J, Boilot JP, Biteau J, et al. Effect of filler size and surface condition of nano-sized silica particles in polysiloxane coatings. Thin Solid Films. 2004;**466**(1–2):114-122

[24] Zhang S, Paik KW. The Effect of Polymer Rebound on Sn-Bi58 Solder ACFs Joints Cracks during a Thermo-Compression Bonding. In: IEEE 67th Electronic Components and Technology Conference (ECTC). IEEE. 2017:2047-2053

[25] Lee K, Kim HJ, Yim MJ, et al. Ultrasonic bonding using anisotropic conductive films (ACFs) for flip chip interconnection. IEEE Transactions on Electronics Packaging Manufacturing. 2009;**32**(4):241-247

[26] Lee K, Oh S, Saarinen IJ, et al. High-speed flex-on-board assembly method using anisotropic conductive films (ACFs) combined with room temperature ultrasonic (US) bonding for high-density module interconnection in mobile phones. Electronic Components and Technology Conference (ECTC). IEEE 61st. IEEE, 2011:530-536

[27] Lee S-H et al. Study on fine pitch flex-on-flex assembly using nanofiber/ solder anisotropic conductive film and ultrasonic bonding method. IEEE Transactions on Components, Packaging and Manufacturing Technology. 2012;**2**.12:2108-2114

[28] Sadeghinia M, Jansen KMB, Ernst LJ. Characterization and modeling the thermo-mechanical cure-dependent properties of epoxy molding compound. International Journal of Adhesion and Adhesives. 2012;**32**:82-88

[29] Suganuma K. Advances in lead-free electronics soldering. Current Opinion in Solid State and Materials Science. 2001;**5**(1):55-64

[30] Kang SK, Sarkhel AK. Lead (Pb)-free solders for electronic packaging. Journal of Electronic Materials. 1994; **23**(8):701-707

[31] Skoog DA, Holler FJ, Crouch SR.
Principles of instrumental analysis.
Cengage learning. 2017

[32] Myung W-R, Kim Y, Jung S-B.
Mechanical property of the epoxy-
contained Sn-58Bi solder with OSP
surface finish. Journal of Alloys and
Compounds. 2014;**615**:S411-S417

[33] Crosby AJ et al. Deformation and
failure modes of adhesively bonded
elastic layers. Journal of Applied
Physics. 2000;**88**(5):2956-2966.
CrossRef. Web

[34] Ferry JD, Myers HS. Viscoelastic
properties of polymers. Journal of the
Electrochemical Society. 1961;**108**(7):
142C. DOI: 10.1149/1.2428174

[35] Chen L et al. The effects of underfill
and its material models on
thermomechanical behaviors of a flip
chip package. IEEE Transactions on
Advanced Packaging. 2001;**24**(1):17-24.
CrossRef. Web